Advanced Research and Real-World Applications of Industry 5.0

Mahmoud Numan Bakkar
Institute of Applied Technology, Abu Dhabi Vocational Education and Training Institute, UAE

Elspeth McKay
Cogniware.com.au, Australia

A volume in the Advances in Logistics, Operations, and Management Science (ALOMS) Book Series

Published in the United States of America by
    IGI Global
    Business Science Reference (an imprint of IGI Global)
    701 E. Chocolate Avenue
    Hershey PA, USA 17033
    Tel: 717-533-8845
    Fax: 717-533-8661
    E-mail: cust@igi-global.com
    Web site: http://www.igi-global.com

Library of Congress Cataloging-in-Publication Data

Names: Bakkar, Mahmoud, 1980- editor. I McKay, Elspeth, 1965- editor.
Title: Advanced research and real-world applications of industry 5.0 /
    Mahmoud Bakkar, and Elspeth McKay, editors.
Description: Hershey : Business Science Reference, 2022. I Includes
    bibliographical references and index. I Summary: "This book is for
    readers who want to improve their understanding of industry 5, including
    the state of the art of real-world applications in industry 5.0,
    highlighting the impact and the risks of countermeasure of industry 4
    has on human life that leads to industry 5"-- Provided by publisher.
Identifiers: LCCN 2022033966 (print) I LCCN 2022033967 (ebook) I ISBN
    9781799888055 (h/c) I ISBN 9781799888062 (s/c) I ISBN 9781799888079
    (ebook)
Subjects: LCSH: Industry 4.0.
Classification: LCC T59.6 .A38 2022 (print) I LCC T59.6 (ebook) I DDC
    303.48/3--dc23/eng/20221017
LC record available at https://lccn.loc.gov/2022033966
LC ebook record available at https://lccn.loc.gov/2022033967

This book is published in the IGI Global book series Advances in Logistics, Operations, and
Management Science (ALOMS) (ISSN: 2327-350X; eISSN: 2327-3518)

British Cataloguing in Publication Data
A Cataloguing in Publication record for this book is available from the British Library.

For electronic access to this publication, please contact: eresources@igi-global.com.

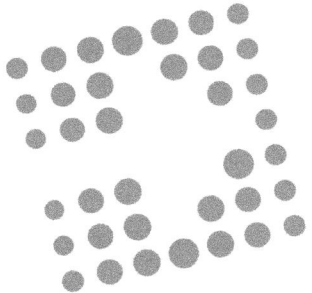

Advances in Logistics, Operations, and Management Science (ALOMS) Book Series

John Wang
Montclair State University, USA

ISSN:2327-350X
EISSN:2327-3518

MISSION

Operations research and management science continue to influence business processes, administration, and management information systems, particularly in covering the application methods for decision-making processes. New case studies and applications on management science, operations management, social sciences, and other behavioral sciences have been incorporated into business and organizations real-world objectives.

The **Advances in Logistics, Operations, and Management Science** (ALOMS) Book Series provides a collection of reference publications on the current trends, applications, theories, and practices in the management science field. Providing relevant and current research, this series and its individual publications would be useful for academics, researchers, scholars, and practitioners interested in improving decision making models and business functions.

COVERAGE

- Networks
- Computing and information technologies
- Risk Management
- Information Management
- Operations Management
- Political Science
- Production Management
- Marketing engineering
- Decision analysis and decision support
- Finance

IGI Global is currently accepting manuscripts for publication within this series. To submit a proposal for a volume in this series, please contact our Acquisition Editors at Acquisitions@igi-global.com or visit: http://www.igi-global.com/publish/.

Titles in this Series

For a list of additional titles in this series, please visit: *http://www.igi-global.com/book-series/*

IGI Global
PUBLISHER of TIMELY KNOWLEDGE

701 East Chocolate Avenue, Hershey, PA 17033, USA
Tel: 717-533-8845 x100 • Fax: 717-533-8661
E-Mail: cust@igi-global.com • www.igi-global.com

This book is dedicated to my mother, Sabta Tarif Dalki; my father, Numan Bakkar; my lovely wife, Hanane; and my family brother and sisters, Mohamed, Zakia, and Mona.

Table of Contents

Section 1
Industry 5.0: Optimizing Globalization and Entrepreneurship

> *Reymond Voutier, eNotus International Inc., USA*
> *Mahmoud Numan Bakkar, Institute of Applied Technology, Abu Dhabi*
>   *Vocational Education and Training Institute, UAE*
> *Guillermo Pivetta, eNotus International Inc., Argentina*

> *Mohammad Izzuddin Mohammed Jamil, Universiti Brunei Darussalam,*
>   *Brunei*

Section 2
Industry 5.0: CyberSecurity Essentials

> *Mahmoud Numan Bakkar, Institute of Applied Technology, Abu Dhabi*
>   *Vocational Education and Training Institute, UAE*

Section 3
Industry 5.0: Utilizing Personalized Design

Section 4
Industry 5.0: Transforming Learning to Education 5.0

Section 5
Industry 5.0: Transforming the Healthcare Industry

Section 6
Industry 5.0: Optimising Industry 4.0 Technologies

Detailed Table of Contents

Section 1
Industry 5.0: Optimizing Globalization and Entrepreneurship

Chapter 1
> *Reymond Voutier, eNotus International Inc., USA*
> *Mahmoud Numan Bakkar, Institute of Applied Technology, Abu Dhabi*
> *   Vocational Education and Training Institute, UAE*
> *Guillermo Pivetta, eNotus International Inc., Argentina*

The chapter is based on the white paper distributed to the group of twenty (G20) and other global organizations concerning the global trust registry plus (GTR +) proposals for a Global Cities Social Venture Fund and affiliated commercial working capital services. This will assist recovery and generate new jobs following the Covid-19 Pandemic. The chapter explains the importance of trust measurement and proposes a global framework for trust calibration and evaluation. It recommends that the trust measure and the scoring register will help create more jobs by improving the efficiency and performance of all small and medium-sized enterprises. In addition, it will help organizations to collaborate and fill the gaps in globalization and communication between nations.

The definition and concepts behind globalization are sometimes confusing for the masses, and studies on globalization often focus on the macroeconomic level. Globalization is the interdependence of people and businesses everywhere that leads to global cultural, political, and economic integration. The purpose of this chapter is to provide background and insight on globalization, from concepts to linking globalization to micro, small, and medium-sized enterprises (MSMEs). Also, to showcase how favorable globalization conditions are beneficial for the growth of MSMEs, which can ultimately result in them expanding their area of operations in terms of sales, number of employees, and size. This chapter presents George Yip's broader framework of globalization that highlights the conditions needed for MSMEs to grow and flourish, including external factors. A framework is proposed to highlight the summarized framework of the external factors and conditions of globalization that influence the growth of MSMEs.

Section 2
Industry 5.0: CyberSecurity Essentials

Currently, hacking threats have increased exponentially because of the massive integration of technology into our daily life practices. Hackers are usually known for their advanced programming skills. They utilize these skills in challenging old systems and work on breaking them to test their capabilities and achieve their desire or motivation. The terminology of cybercrime evolved with the current industry's 4.0 and 5.0 revolutions and the changing of cybersecurity domains. This book chapter will discuss the different types of attacks in the industry 5.0 Era. Show examples of industrial cybersecurity attacks, Industry 5.0 cybersecurity vulnerabilities, and issues.

Cybersecurity is the act of protecting networks, programs, and systems against various hostile and digital assaults. Subset of a security program, it defends cyberspace from escalating assaults and dangers that result in significant damage to resources like finances, information, and applications. Hackers are increasingly targeting firms in the financial and industrial sectors, particularly for the purpose of stealing sensitive data. As a result, many business leaders are turning to cybersecurity to meet their company's security demands and prevent its precious assets from falling into the wrong hands. When it comes to safeguarding software and hardware against unauthorized access and intrusion, cyber assaults play a critical role. The security measure makes use of several security approaches, such as cybersecurity software, access control systems, antivirus and malware security programs, firewalls, and program upgrades, among all system users.

Section 3
Industry 5.0: Utilizing Personalized Design

Chapter 5

Elspeth McKay, Cogniware.com.au, Australia
Mahmoud Numan Bakkar, Institute of Applied Technology, Abu Dhabi
Vocational Education and Training Institute, UAE

The digital economy is now upon us, providing new customer-centric business models fostered during the digital transformation of Industry 5.0. Social networks, mobile applications, data analytics, cloud dependency, and the (IoT) Internet of Things were just vehicles for change. Swept up in this rush for automated business operations is awareness of balancing network complexity and personalizing information measurement systems. This chapter takes a five-pronged approach to situate where the concept of personalized measurement design fits within Industry 5.0. The first section explains how digital transformation has merged multidisciplinary specializations. The second section deals with the preparations for personalized measurement design to enhance the flexibility of online assessment practice. The third section shows how social science knowledge society opens an Industry 5.0 pathway to achieve fully automated human-computer interaction (HCI). The fourth section is about designing flexible online assessments. The final section discusses the next generation of learning analytics.

Chapter 6

Priyadarsini Patnaik, Birla Global University, India
Parameswar Nayak, Birla Global University, India
Siddharth Misra, Birla Global University, India

The transition from Industry 4.0 to Industry 5.0 started when personalization options became available to customers. This revolution aims to bring back the human touch with the convergence of advanced technology towards a degree of personalization to meet the demand of the customers. In the era of Industry 5.0, consumers want to differentiate themselves as unique, and personalized products allow them to express themselves as individuals. This has prompted personal recommendations to become more popular. Despite the increasing popularity of personalized recommendations, little research has been conducted on the impact of these recommendations on user satisfaction. As a result, an online survey was conducted to test the relationships between personalized product recommendations and user satisfaction and proposed a conceptual model. The findings of the study indicated a positive association between personalized product recommendations and consumer satisfaction and highlighted several managerial and practical implications that academics and retailers may find useful.

Section 4
Industry 5.0: Transforming Learning to Education 5.0

Chapter 7

Mahmoud Numan Bakkar, Institute of Applied Technology, Abu Dhabi Vocational Education and Training Institute, UAE
Arshia Kaul, Carpediem EdPsych Consultancy LLP, India

This chapter addresses the relationship between Industry 5.0 and Education 5.0, outlines the future skills required for Industry 5.0, and illustrates how these two sectors interact with the Sustainable Development Goals (SDG) for modernizing Society 5.0. It places a strong emphasis on high-quality education and uses tailored learning to help employees and companies optimize for Industry 5.0. Additionally, it discusses the OECD Learning Compass 2030 and Future of Education and Skills 2030 reports, as well as the International Organization for Standardization (ISO) 21001:2018-ISO standard utilized for future educational management. The chapter emphasizes the need to improve the educational system while presenting various options for adopting personalized education.

Section 5
Industry 5.0: Transforming the Healthcare Industry

Chapter 8

Rita Komalasari, Yarsi University, Indonesia

This chapter aims to present pertinent theoretical frameworks and the most recent results of empirical research in health. It is written for amateurs and experts who wish to understand the industry better. A meta-ethnographic synthesis of research on healthcare applications for Industry 5.0 forms the basis of the review's method. The results show that the link between Industry 5.0 and healthcare applications have a positive relationship. Big data may be used by Industry 5.0 to learn new things and produce symmetrical innovation. It also establishes a digital knowledge network that offers accurate medical data and vital patient records. It establishes a digital knowledge network that offers accurate medical data and vital patient records. This chapter contributes to a better understanding of the link between Industry 5.0 and healthcare applications.

Section 6
Industry 5.0: Optimising Industry 4.0 Technologies

Chapter 9

  Sunil Gupta, UPES University, Dehradun, India
  Monit Kapoor, Chitkara University Institute of Engineering and
    Technology, Chitkara University, Rajpura, India
  Hitesh Kumar Sharma, UPES University, Dehradun, India

Industry 5.0 with the internet of things (IoT) gives a human touch to Industry 4.0 for the development to provide efficiency with automation using robots and machines. The advancement in artificial intelligence makes robots with attached brains like human minds. This brain-machine interface introduces the concept of Industry 5.0, where robots with IoT features tangle with humans and try to work as an agent instead of participants. This digital transformation provides opportunities and challenges. This chapter provides development by various IoT and artificial intelligence-based industries and researchers for the use of Industry 5.0 with their application. In addition, the chapter describes how IoT robots and human values collaborate to give input to Industry 5.0. Finally, the impact of Industrial 5.0 is deliberated based on the economy and manufacturing process with increased productivity.

Chapter 10

  G. Prasad, Chandigarh University, India

The internet of unmanned aerial vehicles (IOU) is a layered network control architecture that is designed primarily for coordinating unmanned aerial vehicle access to controlled airspace and providing navigation with the latest innovative technology upgrades. Human-robot co-working is an emerging subject in Industry 5.0 visions. The IOU provides generic services for a wide range of such as package delivery, traffic monitoring, search and rescue, and multiple applications. In this chapter, the authors present a conceptual model of how such an architecture can be developed as well as the components of an internet of drones system based on artificial intelligence, machine learning, and digital twin are discussed. The future of drones will focus on the thrust area of computer-based domains.

Many people in the world face travel issues in their daily lives, such as traffic congestion, lack of parking spaces, fuel, waste, and pollution. For example, if there are not enough parking spaces on a university campus, it will take a significant amount of time for the students as well as faculty to find parking. In addition, if parking is available outside, employees will have to travel a long distance to enter the building, which will require additional time. Each day, many people in the world must travel a considerable distance to get to their destination. The purpose of this chapter is to resolve the travel issue by proposing a carpool solution using machine learning techniques.

Foreword

Here we are, dear reader, with a book on the fifth revolution in industrial technology, a.k.a., Industry 5.0. Wasn't it only around eight years ago when the then-latest revolution in industrial technology, Industry 4.0, was popularized? Are the two revolutions close together because they are meant to be taken as a set? No; the two revolutions are independent and driven by different needs, productivity through connectivity for 4.0 and sustainability for 5.0. Did the fourth revolution fail? No; Industry 4.0 is a revolution in full swing, while Industry 5.0 is still a nascent one.

So, let's get a clearer picture of the content and differences between the two revolutions. The key concept of Industry 4.0 is the incorporation of AI, Big Data analytics, the Internet of Things, and robotization at the core of manufacturing and supply chains. While Industry 4.0 is about the digital transformation of industry, Industry 5.0 is about "augmenting that digital transformation with a more meaningful and efficient collaboration between humans and the machines and systems within their digital ecosystem" (according to a leading software company) and puts "research and innovation at the service of the transition to a sustainable, human-centric and resilient European industry" (according to the European Union). In my opinion, a gross simplification of the two revolutions is, Industrial 4.0 represents the colonization of the digital space, while 5.0 represents the well-being and sustainment of the colonists.

If you are in a managerial or professional position in a company that either 1) has a reputation of being at the forefront of technology application, you will have many technical questions that you will expect this book to answer; or 2) is presently operating in a low technology mode, but is readying to upgrade in the near term, you will be interested in the planning and conduct of such an activity; or 3) has already undergone a significant upgrade of its technology, and is presently in the grip of post-upgrade doubt and fatigue, you will be wondering if these industrial revolutions might not be so frequent to allow recently-upgraded companies to catch their breath. I can share insights from my experience in several technology developments and deployments that you may find useful in addressing any of these concerns. Hence, let's go through them in order.

Regarding the first concern, I was fortunate enough to get to know Dr. Mahmoud Bakkar—the lead editor of this book and a co-author of four of its chapters—during my stint at the UAE's federal university. Since then, our association has grown in both academics and business. His credentials are impressive because he is tireless, dedicated, and consummate IT professional. Therefore, I was ecstatic when he asked me to contribute a foreword to his upcoming book. Embedding AI into industrial organizations and their business practices was always his passion; much of his life's work has been pursuing this vision. I knew that his writing or editing of a book on this subject was only a matter of time. From previewing the chapters, I could tell Mahmoud has assembled a strong team of like-minded enthusiasts and experts in this field. Before this foreword, I submitted a formal review of the education chapter in the book; I was extremely pleased with the content's comprehensiveness, detail, and clarity. Ergo, considering all of this, I'm highly confident that you, dear reader, will find this book to be a cornucopia of crucial insights for maintaining your company's dominant technological reputation.

To the second concern, many companies acquire their technologies infrequently or sparingly, opting to wait for the right confluence of organizational re-engineering, market forces, and return on investment. Under this strategy, the makeup of their technology stockpile will not align with the orderly and sequenced stages of Industry, e.g., 1.0, 2.0, etc. Does this non-alignment earmark a substandard technological profile? In other words, is the *optimal* structure of technology that follows the content and sequence of the industrial revolutions? The answer is "no," and it stems from how the knowledge that enables technological advancement comes about. Knowledge is rarely developed according to humanity's "master plan," and never preconceived; instead, it is more than not discovered by chance. This means that new technologies from such new knowledge will not by design be tied to older technologies in the same domain, nor are they taxonomically similar to other new technologies of a similar age. So, each stage of the industrial revolution is simply a snapshot of the flagship technologies available when the stage is registered with the public. It is informative like a map: it shows the "lay of the land," and it's up to each company to chart its path using that information.

To the third concern, unless there is an apocalypse that drives humanity back into the stone age, the trend of technological growth at an accelerated pace will continue unabated. Again, this stems from the knowledge that enables technological advancement; it has an exponential growth rate. Although there are different expressions for the growth rate, there is no disagreement that it is exponential. I will reference the growth rate postulated by a well-respected futurist, R. Buckminster Fuller. In his book "Critical Path" (1981), he theorized that our knowledge growth on the planet followed what he dubbed the "doubling curve." Starting with all the knowledge thought to have been acquired for the three million years up to year one

CE, Fuller claimed that it doubled by 1500 CE, then doubled again by 1750 CE, and then doubled a third time by 1900 CE. Within the next 45 years, spurred by two World Wars, knowledge again doubled. By 1975, it doubled again. Since the "Critical Path" was published, other futurists have continued to update the doubling curve. With the advent of affordable home computers, the doubling curve shortened to 24 months; by the turn of the new millennium, it would be 18 months. So here we are two decades into the 21st century, with the doubling of knowledge estimated at less than 12 months. Of course, the evolution of industrial technology is moderated by the availability of appropriate resources, so it is slower by magnitudes than the knowledge doubling curve. Notwithstanding, it is a certainty that the growth of industrial technology will be unrelenting. No matter how leading-edge the technology recently purchased is, it is inevitable that newer and seemingly better technology will be just on the horizon.

Personally, this book has provided me with useful information and further substantiation of my ideas concerning online learning and higher education possibilities in the metaverse. In my mind, these ideas associate nicely with Industry 5.0's topics of human-centric-ness, resilience, and sustainability. I'm sure company managers will make similar connections with their business organizations and processes. Aside from the fact that many companies are still engaged in earlier industrial revolutions, the themes of Industry 5.0 are nonetheless gravely topical because of the social tension and ecology side-effects that many industrial technologies give rise to. Once more companies expand their business priorities to embrace the human-centric-ness, resilience, and sustainability contained in Industry 5.0, the more vibrant and productive industry will become. I wish Dr Mahmoud Bakkar, co-editor Dr Elspeth McKay, and their team of contributing experts every success in promulgating that message to the readers.

Kevin Anthony Jones
Singapore Institute of Technology, Singapore

Preface

OVERVIEW

The book overviews Industry 5.0 and related advanced research and real-world applications. The respective chapters identify the conceptual framework of Industry 5.0, the Industry 5.0 technology roadmap, digital transformation and Industry 5.0, the role of artificial intelligence and its impact on humans, and discuss the main technology drivers for Industry 5.0. The book presents performance metrics to assist the manufacturer in identifying the risk factors associated with deploying Industry 5.0 technology. They also discuss real-time issues, problems, and applications with corresponding solutions and suggestions; yielding new theoretical conclusions, tools, and techniques for Industry 5.0 and comprising theoretical and real-world application approaches. The book advances as a worthy asset for novices and professionals in the field.

There can be no doubt that awareness of the digital economy is prominent for maintaining successful business operations. In the recent past, the fourth Industrial Revolution (commonly known as Industry 4.0) focused on sustaining the industrial sector. This digital transformation radically changed many business operations concerning product development, service provision, and processes that changed business operations modelling (Hahn, 2020). Yet, the emerging awareness of smoother automation saw the arrival of Industry 5.0. It brings forward digital product integration, increasing processes and business operational efficiency (Elangovan, 2021). Industry 5.0 deals with the duality of human and digital machine intelligence through the combined efforts of the industrial Internet of Things (IoT) and artificial intelligence (AI). According to Elangovan (2021), the digital technological importance of Industry 5.0 is the ability to utilize collaborative robotics, known as co-robots, to speed up manufacturing profitability. In addition, there are many other sectors where Industry 5.0 provides a seamless improvement of service delivery. The global nature of this book's chapter authorship is a testimony to the widespread awareness of adopting the visionary spirit offered to humanity by Industry 5.0.

OBJECTIVES OF THE BOOK

This book provides relevant theoretical frameworks and the latest empirical research findings. It's written for novices and professionals who want to improve their understanding of Industry 5.0. Presenting the state-of-the-art real-world applications in Industry 5.0 highlights the impact and the risks' countermeasure of Industry 4.0 on human life that led to Industry 5.0. It also introduces the potential of Industry 5.0 to advance the digital workplace. The chapters focus on emerging Industry 5.0 best practice in corporate performance that applies to business, the socialization of healthcare, education, and commonplace daily living. In so doing, it brings forward the transformation of traditional Industry 4.0 frameworks that led the push into the Industry 5.0 environmental change, involving Internet interactivity that has succeeded in the business arena in a language familiar to teaching and learning institutions in schools and higher education.

TARGET AUDIENCE

This book's target audience will be professionals and researchers working in all educational fields. Industry 5.0 is a multidisciplinary research area. Engineering and Information Technology researchers will greatly benefit from reading this book. Business computing-oriented studies will be able to integrate their studies with this book's topics, further to many other disciplines, such as medicine and biomedical science. Moreover, the book will provide insights and support to executives who manage manufacturing, knowledge information, and organizational development in different work communities and environments. Other audiences for this book include industry training developers, corporate trainers, courseware designers, government sector specialists, infrastructure policymakers, educational technology practitioners and researchers, school teachers and administrators, higher education institutions, and postgraduate students.

Section 1: Industry 5.0 – Optimizing Globalization and Entrepreneurship

In the manufacturing industry, Industry 5.0 represents the next phase in its development, integrating advanced technologies, including artificial intelligence and blending human-centric values like empathy, creativity, and social responsibility with IoT and robotics (OpenAI, 2023). The goal of Industry 5.0 is to optimize globalization and entrepreneurship, using advanced technology and human values to create new opportunities for businesses to thrive.

Here are some of the key factors enabling Industry 5.0 to Optimize Globalization and Entrepreneurship:

- **Collaboration**: Industry 5.0 enhances collaboration between humans and machines and businesses and their stakeholders, like suppliers, customers, and partners. This results in new networks that allow companies to leverage each other's expertise and strengths, enabling them to innovate more quickly and efficiently (OpenAI, 2023);
- **Customized Products**: Businesses can now offer more customized products and services by using advanced technologies to meet their customers' unique needs and preferences. Additionally, it allows companies to differentiate themselves in a crowded marketplace and improve customer satisfaction (OpenAI, 2023);
- **Optimizing Operations With Industry 5.0:** Businesses can optimize their operations through automated repetitive tasks, real-time analytics of data, and decision-making based on insights generated by AI and other advanced technologies. The result is an increase in productivity, a reduction in costs, and an increase in profitability (OpenAI, 2023); and
- **Emphasizing Social Responsibility:** Industry 5.0 strongly emphasises social responsibility and sustainability to reduce their environmental impact, promote diversity and inclusion, and contribute to the well-being of communities in which they operate. In other words, businesses expect to take a more proactive approach (OpenAI, 2023).

In summary, Industry 5.0 is about leveraging advanced technologies to optimize globalization and entrepreneurship while promoting human values such as collaboration, customization, efficiency, and social responsibility. By doing so, businesses can create new opportunities for growth and innovation in the global marketplace while contributing to a more sustainable and fair future.

Section 2: Industry 5.0 – Cybersecurity Essentials

Industry 5.0 provides companies with enormous opportunities to grow in the global market, but it also raises major cybersecurity challenges to ensure the safety of people, assets, and data.

Here are some of the key cybersecurity essentials that businesses should consider in Industry 5.0:

- **Secure Devices and Networks:** In Industry 5.0, devices and networks are more interconnected than ever before. This circumstance makes them vulnerable to cyber threats such as hacking, malware, and data breaches. To ensure the security of devices and networks, businesses should implement robust cybersecurity protocols that include firewalls, encryption, multifactor authentication, and regular security updates (OpenAI, 2023);

- **Data Protection:** Industry 5.0 generates massive amounts of data that must be protected from unauthorized access and theft. Businesses should implement data protection measures such as data encryption, access control, and data backup and recovery to ensure that their data remains secure and available during a cyber-attack or other disruption (OpenAI, 2023);

- **Risk Management:** Industry 5.0 involves significant risks, including cybersecurity risks. Businesses should conduct regular risk assessments to identify potential threats and vulnerabilities and develop risk management plans, including incident response, disaster recovery, and business continuity measures (OpenAI, 2023);

- **Employee Training:** Industry 5.0 requires a highly skilled workforce trained in the latest cybersecurity practices and protocols. Businesses should provide ongoing training and education to employees to ensure they know the latest cyber threats and how to prevent and respond to them (OpenAI, 2023); and

- **Collaboration:** Industry 5.0 requires collaboration between businesses, governments, and other stakeholders to address cybersecurity challenges. Businesses should work together with their partners, suppliers, and customers to develop cybersecurity policies and protocols that protect the entire ecosystem (OpenAI, 2023).

In summary, cybersecurity is an essential component of Industry 5.0 that must be addressed to ensure people's assets, data safety and security. By implementing robust cybersecurity protocols, businesses can leverage the benefits of Industry 5.0 while protecting themselves and their stakeholders from cyber threats.

Section 3: Industry 5.0 – Utilizing Personalized Design

One of the key benefits of Industry 5.0 is the ability to utilize personalized design to create products and services tailored to individual customers' specific needs and preferences.

Here are some ways in which Industry 5.0 can utilize personalized design:

- **Customized Products:** With advanced technologies such as 3D printing and AI, businesses can create customized products that meet individual customers' unique needs and preferences. For example, a shoe manufacturer can use 3D printing to create shoes tailored to the exact size and shape of a customer's feet (OpenAI, 2023);

- **Personalized Experiences:** Industry 5.0, businesses can use data and AI to create personalized customer experiences. For example, a streaming service can use AI to recommend movies and TV shows based on a customer's viewing history and preferences (OpenAI, 2023);
- **Co-Creation:** Industry 5.0, businesses can collaborate with customers to co-create products and services. For example, a car manufacturer can use virtual reality technology to allow customers to design their cars and provide feedback on the design process (OpenAI, 2023); and
- **Mass Customization:** Industry 5.0, businesses can leverage mass customization to create products customized to many customer profiles. For example, a cosmetics company can use AI to analyze customer data and generate a range of makeup shades tailored to a diverse customer base's skin tones and preferences (OpenAI, 2023).

By utilizing personalized design in Industry 5.0, businesses can create products and services tailored to individual customers' specific needs and preferences, increasing customer satisfaction, loyalty, and revenue. Personalized design also allows firms to differentiate themselves in a crowded marketplace and to create new opportunities for growth and innovation.

Many instances of personalized data collection techniques operate within the industry 5.0 model. These contributions involve training and machine learning projects; one of the biggest challenges is obtaining enough data sets to train algorithms. However, it is unwise to take humans out of the process completely; instead, they combine human workers with AI background skills. The first job identifies how best to split the data into as many components as possible to automate each piece one at a time. Then the next step involves human labeling the data; AI learns over time and gradually takes over (Miller, 2019).

At this point, it is prudent to consider the human dimension of human-computer interaction (HCI). The human dimensions of HCI are a major part of the complex digital usability framework. Two dimensions have distinct and quite separate contexts. One relates to the social context of computing, while the other deals with the machine dimension, where the technical aspects deal with the system's performance of the technical components (McKay, 2008). Nevertheless, it remains incumbent upon system designers to be aware of HCI, involved in the automated functionality of most Industry 5.0 environments.

Section 4: Industry 5.0 – Transforming Learning to Education 5.0

Education 5.0 is an evolution of traditional education models incorporating the latest technological advances and human-centric values to provide a more personalized, collaborative, and experiential learning experience.

Here are some learning strategies of how Industry 5.0 is transforming learning into Education 5.0:

- **Personalized Learning:** In Education 5.0, students are empowered to learn at their own pace and according to their own learning style. Technology such as AI can create personalized learning paths for students based on their strengths, weaknesses, and interests (OpenAI, 2023);
- **Collaborative Learning:** Education 5.0 emphasizes collaboration and teamwork, mirroring real-world work environments. Technology such as virtual and augmented reality can create immersive learning experiences that promote collaboration and teamwork among students (OpenAI, 2023);
- **Experiential Learning:** In Education 5.0, students learn by doing rather than simply memorizing information. Technology such as simulations, virtual labs, and gamification can create experiential learning opportunities that engage students and help them retain knowledge (OpenAI, 2023);
- **Lifelong Learning:** Education 5.0 recognizes that learning is a lifelong process and not just something that happens during a person's formal education years. Technology such as e-learning platforms and online courses make it possible for individuals to continue learning and upskilling throughout their lives (OpenAI, 2023); and
- **Student-Centered Education:** Education 5.0 places the student at the center of the learning experience rather than the teacher. This activity means that students are encouraged to take ownership of their learning and to actively engage in the learning process (OpenAI, 2023).

By transforming learning to Education 5.0, we can create a more personalized, collaborative, and experiential learning experience that prepares students for success in the 21st-century workforce. The integration of advanced technologies such as AI, virtual and augmented reality, and e-learning platforms, combined with human-centric values such as empathy, creativity, and social responsibility, makes Education 5.0 a powerful tool for unlocking the potential of every student.

Section 5: Industry 5.0 – Transforming the Healthcare Industry

Industry 5.0, which combines advanced technologies such as artificial intelligence (AI), robotics, and the Internet of Things (IoT) with human-centric values, is transforming the healthcare industry. Here are some ways in which Industry 5.0 is transforming healthcare:

- **Personalized Medicine:** Industry 5.0, healthcare is becoming more personalized, with treatments tailored to individual patients' specific needs and characteristics. AI and machine learning used to analyze large amounts of patient data to identify patterns and develop personalized treatment plans (OpenAI, 2023);
- **Remote Patient Monitoring:** The IoT enables remote patient monitoring, allowing doctors and other healthcare professionals to track patients' health remotely and intervene as needed. Wearable devices and other IoT-enabled devices can provide real-time data on a patient's vital signs, activity level, and other health metrics (OpenAI, 2023);
- **Robotics-Assisted Surgery:** Robotics plays an increasingly important role in surgery, with robots performing minimally invasive procedures with greater precision and accuracy. This activity can result in shorter recovery times, fewer complications, and better patient outcomes (OpenAI, 2023);
- **Telemedicine**: Becoming more widespread, enabling patients to consult with healthcare professionals remotely using video conferencing and other communication technologies. This activity can particularly benefit patients living in remote areas or with mobility issues (OpenAI, 2023);
- **Predictive Analytics:** Predictive analytics, enabled by AI and machine learning, is being used to identify patients who are at risk of developing certain conditions, allowing healthcare providers to intervene before the situation becomes more serious (OpenAI, 2023); and
- **Health Data Management:** Health data management is becoming increasingly important, with advanced technologies such as blockchain being used to ensure the security and privacy of patient data (OpenAI, 2023).

By transforming healthcare with Industry 5.0 technologies and human-centric values, we can improve patient outcomes, reduce costs, and increase access to care. Integrating AI, robotics, and the IoT with human-centric values such as empathy, compassion, and social responsibility can lead to a more patient-centered approach to healthcare, where individual needs are prioritized over one-size-fits-all solutions.

Section 6: Industry 5.0 – Optimizing Industry 4.0 Technologies

Industry 5.0 builds upon Industry 4.0 technologies and takes them to the next level by integrating them with human-centric values. While Industry 4.0 focused on the integration of automation, data exchange, and digital technologies in the manufacturing process, Industry 5.0 takes a more holistic approach, incorporating advanced technologies such as artificial intelligence (AI), robotics, and the Internet of Things (IoT) with human-centric values to optimize the benefits of Industry 4.0 technologies (OpenAI, 2023). Here are some ways in which Industry 5.0 is optimizing Industry 4.0 technologies:

- **Human-Machine Collaboration:** Industry 5.0 recognizes the importance of human expertise in manufacturing and seeks to integrate this with advanced technologies such as robotics and AI to optimize performance. This approach enables humans and machines to work together seamlessly, each contributing unique strengths to manufacturing (OpenAI, 2023);
- **Data-Driven Decision-Making:** Industry 5.0 takes the data generated by Industry 4.0 technologies and uses it to inform decision-making. With AI and machine learning, it is possible to analyze large amounts of data to identify patterns and make predictions about the manufacturing process, enabling manufacturers to optimize production and reduce waste (OpenAI, 2023);
- **Predictive Maintenance:** The IoT can monitor machines and equipment in real-time, identifying potential issues before they become serious problems. This activity can help to reduce downtime and increase productivity, improving overall efficiency in the manufacturing process (OpenAI, 2023);
- **Intelligent Supply Chain Management:** Industry 5.0 takes a holistic approach to supply chain management, using advanced technologies such as blockchain to create a transparent and secure supply chain. This activity can help to reduce waste, improve efficiency, and ensure that products are manufactured and delivered to customers in a timely and cost-effective manner (OpenAI, 2023); and
- **Autonomous Production:** Industry 5.0 incorporates autonomous production technologies, such as self-driving vehicles and drones, to optimize manufacturing. These technologies can transport materials, perform inspections, and even assemble products, reducing the need for human intervention and increasing efficiency (OpenAI, 2023).

By integrating Industry 4.0 technologies with human-centric values, Industry 5.0 is optimizing the manufacturing process, making it more efficient, sustainable, and cost-effective. In addition, the combination of advanced technologies such as AI, robotics, and the IoT with human expertise and values such as empathy, creativity, and social responsibility can lead to a manufacturing process that is more innovative, collaborative, and effective.

TERMINOLOGY

A short word is given here to afford a more comfortable reading stance, which is necessary as the Industry 5.0 paradigm extends across many philosophical fields. As a result, authors may sometimes refer to the following terms without providing satisfactory clarification. For instance:

- **AI (Artificial Intelligence):** Refers to computer systems simulation of human intelligence processes.
- **HCI (Human-Computer Interaction):** some Authors refer to this term as the human-computer *interface* (instead of interaction). If so, they identify that they describe the screen-based characteristics rather than the interaction.
- **IoT (Internet of Things):** Refers to physical objects linked through wired and wireless networks. It refers to the collection of internet-connected devices that can communicate autonomously over the Internet without needing a human to initiate the communication.
- **ISO (International Organization for Standardization):** Represents the first full-length study of the largest nongovernmental, global regulatory network whose scope and influence rival the UN system (Murphy, 2009).
- **MSMs (Micro, Small and Medium-Sized Enterprises):** This term refers to a major worldwide source of entrepreneurial skills, innovation, and employment (Morales Pedraza, 2021).
- **SDG (Sustainable Development Goals):** Refers to 17 universal goals,169 targets, and 230 indicators leading up to 2030 (Allen et al., 2016).

INDUSTRY 5.0-RELATED AREAS

- Industry 5.0 and How It's Supporting Human Workers.
- Industry 5.0 and Its Optimization Models
- Industry 5.0 and Its Challenges and Opportunities
- Industry 5.0 and Collaborative Robots
- Industry 5.0 and Creative People
- Industry 5.0 and Human Thought and Behaviours
- Industry 5.0 Puts Human Beings at the Center of Industrial Production

- Industry 5.0 and Sustainability
- Industry 5.0 and Personalization Design
- Industry 5.0 and Human Intelligence
- Industry 5.0 and Job Market
- Industry 5.0 and Closing the Loop Between Physics and Design
- Industry 5.0 and Real-Time Automation
- Industry 5.0 and Cybersecurity Threats
- Industry 5.0 and Industry 4.0 Technology Optimization and Utilization Models
- Industry 5.0 and Globalization, Entrepreneurship
- Industry 5.0 and Digital Transformation Strategies
- Industry 5.0 and Strategies

SCHOLARLY VALUE AND CONTRIBUTION

The chapters in this book directly compare traditional Industry 4.0 – 5.0 transformation progress showing emerging business and education sector settings. These examples provide a diverse range of positive possibilities that link information management techniques to enhance the leverage of Industry 5.0 tools in various specialist fields. Through the global nature of the authorship contribution, this book reveals the impact of increased awareness for promoting more effective digital transformation to Industry 5.0 that benchmarks forthcoming action.

BOOK ORGANIZATION AND CONTRIBUTIONS

This book is organized into 11 chapters, which fall into the following themes: "Industry 5.0: Optimizing Globalization and Entrepreneurship"; "Industry 5.0: Cybersecurity Essentials"; "Industry 5.0: Utilizing Personalized Design"; "Industry 5.0: Transforming Learning Into Education 5.0"; "Industry 5.0: Transforming the Healthcare Industry"; and "Industry 5.0: Optimizing Industry 4.0 Technologies."

Section 1: Industry 5.0 – Optimizing Globalization and Entrepreneurship

Industry 5.0 is a crucial need to optimize the effects of globalization, through its collaboration dogma, either between humans and machines or between global businesses, which will facilitate worldwide innovation and know-how tech sharing. Furthermore, having more customized products will open the door for more entrepreneurs to flourish and grow gradually in the global market by differentiating themselves in a crowded marketplace and improving customer satisfaction. Furthermore, industry 5.0 businesses can optimize their operations through automated repetitive tasks and strongly emphasise social responsibility and sustainability through proactive measures for protecting the environment. Therefore, the first chapter discusses the need for a global trust registry, which facilitates international collaboration through global quality measurement and scoring. In addition, the second chapter details the meaning of globalization and proposes a framework for enabling small and medium-sized enterprises (MSMEs) to grow globally.

Chapter 1: GTR + Connecting G20 Vision With Actions in the Industry 5.0 Era

The chapter is based on the White Paper distributed to G20 and other global organizations concerning the Global Trust Registry and proposals for a Global Cities Social Venture Fund and affiliated commercial working capital services to assist recovery and generate new jobs following the Covid Pandemic. The chapter explains the trust measurement's importance and proposes a global framework for trust calibration and evaluation. Furthermore, it proposes that trust measurement will assist in creating more jobs by enhancing the efficiency and performance of all small and medium enterprises. Also, it will help the organizations to collaborate and fill the gaps of globalization and communications between nations.

Chapter 2: Globalization and Entrepreneurship in the Industry 5.0 Era

The definition and concepts behind globalization are sometimes confusing for the masses, and studies on globalization often focus on the macroeconomic level. Globalization is the interdependence of people and businesses everywhere, leading to global cultural, political, and economic integration. This chapter provides background and insight on globalization, from concepts to linking globalization to micro, small, and medium-sized enterprises (MSMEs). Also, to showcase how favorable globalization conditions are beneficial for the growth of MSMEs, which can ultimately result in them expanding their area of operations in terms of sales, number of employees, and size. This chapter presents George Yip's broader framework of globalization, highlighting the conditions needed for MSMEs to grow and flourish, including external factors. A framework is proposed to highlight the summarized framework of the external factors and conditions of globalization that influence the growth of MSMEs.

Section 2: Industry 5.0 – Cybersecurity Essentials

The Internet of Things (IoT) has become prominent with Industry 5.0. This sense of real-time sensing and responding to queries has brought about a real digital transformation. However, along with the rush for data-driven smart government service provision, the literature shows many cybersecurity challenges in IoT require urgent research on policy and usage (Chatfielda & Reddick, 2019). Knowing how to protect our digital data space has become imperative for many of us. We assume that our digital details are kept safe and secure. Yet, significant corporate information breaches have recently put people on guard to adopt stronger protection strategies. In addition, employees now use their mobile phones to conduct their work practices in the business sector. These challenges magnify because cybersecurity compliance and policy concerns are poorly understood (Ameen et al., 2021). In seeking answers to this dilemma, the book's second part communicates how the authors face cybersecurity challenges within an Industry 5.0 lens. Along with the multitude of new-age digital transformation attractions afforded by Industry 4.0 and Industry 5.0 that promote smartification for various everyday life activities, there are new challenges people face in the form of cybersecurity risks (Kumar & Mallipeddi, 2022). Therefore, the two chapters below discuss the cybersecurity essentials and the cyber crime security threats for Industry 5.0.

Chapter 3: Cybersecurity Essentials for Industry 5.0

Hacking threats have increased exponentially because of the massive integration of technology into our daily life practices. Hackers are usually known for their advanced programming skills. They utilize these skills in challenging old systems and work on breaking them to test their capabilities and achieve their desire or motivation. The terminology of cybercrime evolved with the current Industry's 4.0 and 5.0 revolutions and the changing of cybersecurity domains. Examples of industrial cybersecurity essentials typically involve: Industry 5.0 cybersecurity vulnerabilities and issues; types of cybersecurity skills required for the 5.0 Industry Era; protecting secrets and ensuring integrity; industrial approach for cybersecurity in the Industry 4.0 and 5.0 era; information security standards and frameworks; IT and OT concepts; cybersecurity attacks that threaten and countermeasures for industry 5.0.

Chapter 4: Industry 5.0 and Cyber Crime Security Threats

Cybersecurity is the act of protecting networks, programs, and systems against various hostile and digital assaults. As a subset of another security program, it defends cyberspace from escalating assaults and dangers that significantly damage resources like finances, information, and applications. Hackers are increasingly targeting firms in the financial and industrial sectors, particularly to steal sensitive data (Lezzi, Lazoi, & Corallo, 2018). As a result, many business leaders are turning to cybersecurity to meet their company's security demands and prevent its precious assets from falling into the wrong hands. However, when it comes to safeguarding software and hardware against unauthorized access and intrusion, cyber assaults play a critical role. The security measure uses several security approaches, such as cybersecurity software, access control systems, antivirus and malware security programs, firewalls, and program upgrades, among all system users.

Section 3: Industry 5.0 – Utilizing Personalized Design

Industry 5.0 has elevated people's expectations of how to deal with the increased scope of digital disruptions in their everyday lives. This type of digital disruption sets up a communal approach towards exploring and sharing new ideas, techniques that sometimes do not work in particular settings, and the best tools to get the job done (Bhatnagar et al., 2021). The selected chapters for this part represent the wide range of authors' views, including personalized performance measurement, product recommendation and user satisfaction, and the historical account of the effects of the digital transformation from Industry 4.0 to Industry 5.0. Customized or otherwise called here as personalized design, is achieved through: customized products; mass customization using AI-driven data; co-creation between the customers and the business; and personalized experiences analyses using AI-driven data.

Chapter 5: Personalized Measurement Design – Implementing Industry 5.0

The digital economy is now upon us, providing new customer-centric business models fostered during the digital transformation of Industry 5.0. Social networks, mobile applications, data analytics, cloud dependency, and the IoT were just vehicles for change. Swept up in this rush for automated business operations is awareness of balancing network complexity and personalizing information measurement systems. This chapter takes a five-pronged approach to situate where the concept of personalized measurement design fits within Industry 5.0. The first section explains how digital transformation has merged multidisciplinary specializations. The second section deals with the preparations for personalized measurement design to enhance the flexibility of online assessment practice. The third section shows how social science knowledge society opens an Industry 5.0 pathway to achieve fully automated HCI. The fourth section is about designing flexible online assessment practices. The final section discusses the next generation of learning analytics.

Chapter 6: Personalized Product Recommendation and User Satisfaction Reference to Industry 5.0

The shift from Industry 4.0 to Industry 5.0 began when customization options became available to customers. This revolution aims to bring back the human touch with the convergence of advanced technology towards a degree of personalization to meet the customers' demands. In the era of Industry 5.0, consumers want to stand out as unique, where personalized products enable them to express themselves as individuals. This activity has prompted personal recommendations to become more popular. However, despite the popularity of personalized advice, there has been limited research on how these recommendations influence user satisfaction. Hence, an online survey was conducted to test the relationships between customized product recommendations and user satisfaction and proposed a conceptual model. The study's findings indicated a positive association between personalized product recommendations and consumer satisfaction and outlined several managerial and practical implications that academics and retailers may find useful.

Section 4: Industry 5.0 – Transforming Learning Into Education 5.0

Following on from the ideas behind borderless education and dealing with the emergent innovations of educational opportunities brought about by the advent of Industry 5.0 (Doyle-Kent & Shanahan 2022), the following chapter describes how such an andrological educational model is emerging to guide the next generation of automated learning design practices. Industry 4.0 gave rise to notions of a borderless education (Cep Ubad et al., 2020). This leap of faith signalled the need to rethink pedagogy/andragogy learning design. However, educational technologists have been working on Industry 4.0 designs (International Conference on Educational, 2020). Still, in 2023 they have not yet caught up with Industry 5.0, suggesting there are opportunities for mounting effective learning programs designed for an Industry 5.0 rollout. However, important processing steps and structures are necessary for this imaginative stance to be possible. Accordingly, at the Innovation Hub in the University of Zimbabwe, they are implementing such sustainable development goals (SDG) for water, energy, and food, resources which are in critical shortage in Harare, as part of its mandate to implement the newly introduced Education 5.0 (Togo & Gandidzanwa, 2021). Education 5.0 delivery is established using personalized learning, collaborative learning, experiential learning, lifelong learning, and student-centered education strategies (OpenAI, 2023). The following chapter explains the previously mentioned strategies and the needed educational skills for industry 5.0 implementation.

Chapter 7: Education 5.0 Serving Future Skills for Industry 5.0 Era

This chapter addresses the relationship between Industry 5.0 and Education 5.0, outlines the future skills required for Industry 5.0, and illustrates how these two sectors interact with the Sustainable Development Goals (SDG) for modernizing Society 5.0. It strongly emphasizes high-quality education and uses tailored learning to help employees and companies optimize for Industry 5.0. Additionally, it discusses the OECD Learning Compass 2030 and Future of Education and Skills 2030 reports, as well as the International Organization for Standardization (ISO) 21001:2018-ISO standard utilized for future educational management. Finally, the chapter emphasizes the need to improve the educational system while presenting various options for adopting personalized education.

Section 5: Industry 5.0 – Transforming the Healthcare Industry

Industry 5.0 plays a crucial role in transforming the healthcare industry using different strategies such as: personalized medicine; AI-driven medical data; remote patient monitoring; robotics-assisted surgery; telemedicine; predictive analytics; and health data management (OpenAI, 2023). The following chapter discusses the regulatory shift needed to cope with industry 5.0 applications.

Chapter 8: Regulatory Shift – Healthcare Industry 5.0 Applications

This chapter presents pertinent theoretical frameworks and the most recent results of empirical research in the healthcare field. It is written for amateurs and experts who wish to understand the Industry better. A meta-ethnographic synthesis of research on healthcare applications for Industry 5.0 forms the basis of the review's method. The result shows that the link between Industry 5.0 and healthcare applications would have a positive relationship. Big data may be used by Industry 5.0 to learn new things and produce symmetrical innovation. Also, it establishes a digital knowledge network that offers accurate medical data and vital patient records, creating a network that provides accurate medical data and vital patient records. This chapter contributes to a better understanding of the link between Industry 5.0 and healthcare applications.

Section 6: Industry 5.0 – Optimizing Industry 4.0 Technologies

Industry 5.0 builds upon Industry 4.0 technologies and takes them to the next level by integrating them with human-centric values. While Industry 4.0 focused on the integration of automation, data exchange, and digital technologies in the manufacturing process, Industry 5.0 takes a more holistic approach, incorporating advanced technologies such as artificial intelligence (AI), robotics, and the Internet of Things (IoT) with human-centric values to optimize the benefits of Industry 4.0 technologies. The following chapters exemplify the optimization of Industry 4.0 technology using Industry 5.0 strategies.

Chapter 9: Industry 5.0 and the Collaborative Approach of the Internet of Things With Artificial Intelligence

Industry 5.0 with the IoT gives a human touch to Industry 4.0 for the development to provide efficiency with automation using robots and machines. The advancement in artificial intelligence makes robots with attached brains like human minds. This brain-machine interface introduces the concept of Industry 5.0, where robots with IoT features tangle with humans and tries to work as agents instead of participants. This digital transformation provides opportunities and challenges. This chapter includes development by various IoT and artificial intelligence-based industries and research for using Industry 5.0 with their application. In addition, the chapter describes how IoT robots and human values collaborate to give input to Industry 5.0. Finally, the impact of Industrial 5.0 is deliberated based on the economy and manufacturing process with increased productivity.

Chapter 10: Internet of Unmanned Aerial Vehicles in Industry 5.0

The Internet of Unmanned (IOU) devices is a layered network control architecture designed to coordinate unmanned aerial vehicle access to controlled airspace and provide navigation with the latest innovative technology upgrades. Human-robot co-working is an emerging subject in Industry 5.0 visions. These IOU devices provide generic services such as: package delivery; traffic monitoring; search and rescue; and multiple applications. The components of an Internet of Drones (IOD) system based on AI, machine learning, and digital twins are discussed. The future of drones will focus on the thrust area of computer-based domains.

Chapter 11: Carpooling Solutions Using Machine Learning Tools

The major reason for writing this chapter is the author's daily disruptive experiences. For instance: traffic congestion; a lack of parking spaces; fuel waste; and pollution. If there is insufficient parking within a university's campus, looking for parking may take a long time. Moreover, if parking is available outside the university, employees will have to travel a great distance to access their building, which may take longer. For university residential reasons, an identical issue will arise. For example, suppose a university has several research stations that are geographically separate. In that case, communication occurs between them and the main office using office cars, although occasionally leased vehicles are obtained from a travel agency. The cost of transportation will rise as a result. Similarly, all students not from the dormitory and are day scholars travelling from their homes are considered day scholars. Every day, they must drive a significant distance to go to the institution in rented automobiles. This chapter deals with the concept of carpooling using machine learning techniques.

Mahmoud Numan Bakkar
Institute of Applied Technology, Abu Dhabi Vocational Education and Training Institute, UAE

Elspeth McKay
Cogniware.com.au, Australia

REFERENCES

Ameen, N. T. A., Tarhini, A., Shah, M. H., Madichie, N., Paul, J., & Choudrie, J. (2021). Keeping customers' data secure A cross-cultural study of cybersecurity compliance among the Gen-Mobile workforce. *Computers in Human Behavior*, *114*, 106531. doi:10.1016/j.chb.2020.106531

Bhatnagar, V., Sinha, S., Johri, P., & Bali, V. (2021). *Disruptive Technologies for Society 5.0: Exploration of New Ideas, Techniques, and Tools*. CRC Press.

Cep Ubad, A., Vina, A., & Ade Gafar, A. (2020). Borderless Education as a Challenge in the 5.0 Society. In *Proceedings of the 3rd International Conference on Educational Sciences (ICES 2019)*. CRC Press.

Chatfielda & Reddick. (2019). A framework for Internet of Things-enabled smart government: A case of IoTcybersecurity policies and use cases in U.S. federal government. *Government Information Quarterly, 36*(2), 346-357.

Doyle-Kent, M., & Shanahan, B. W. (2022). The development of a novel educational model to successfully upskill technical workers for Industry 5.0: Ireland a case study. *IFAC-PapersOnLine, 55*(39), 425–430. doi:10.1016/j.ifacol.2022.12.072

Elangovan, U. (2021). *Industry 5.0: The Future of the Industrial Economy*. CRC. doi:10.1201/9781003190677

GBD 2015 SDG Collaborators. (2016). Measuring the health-related Sustainable Development Goals in 188 countries: A baseline analysis from the Global Burden of Disease Study 2015. *Lancet, 388*(10053), 1813–1850. doi:10.1016/S0140-6736(16)31467-2

Hahn, G. J. (2020). Industry 4.0: A supply chain innovation perspective. *International Journal of Production Research, 58*(5), 1425–1441. doi:10.1080/00207543.2019.1641642

International Conference on Educational. (2020). Borderless education as a challenge in the 5.0 society. In *proceedings of the 3rd International Conference on Educational Sciences (ICES 2019).* Routledge.

Kumar, S., & Mallipeddi, R. R. (2022). Impact of cybersecurity on operations and supply chain management: Emerging trends and future research directions. *Production and Operations Management, 31*(12), 4488–4500. doi:10.1111/poms.13859

Lezzi, M., Lazoi, M., & Corallo, A. (2018). Cybersecurity for Industry 4.0 in the current literature: A reference framework. Computers. *Computers in Industry, 103,* 97–110. doi:10.1016/j.compind.2018.09.004

McKay, E. (2008). *The Human-Dimensions of Human-Computer Interaction: Balancing the HCI Equation.* IOS Press.

Miller, R. (2019). *Superb AI generates customized training data for machine learning projects.* TechCrunch.

Morales Pedraza, J. (2021). The Micro, Small, and Medium-Sized Enterprises and Its Role in the Economic Development of a Country. *Business and Management Research, 10*(1), 33–44. doi:10.5430/bmr.v10n1p33

Murphy, C. (2009). *The International Organization for Standardization (ISO) global governance through voluntary consensus.* Routledge. doi:10.4324/9780203884348

OpenAI. (2023). *ChatGPT.* https://chat.openai.com/chat

Togo, M., & Gandidzanwa, C. P. (2021). The role of Education 5.0 in accelerating the implementation of SDGs and challenges encountered at the University of Zimbabwe. *International Journal of Sustainability in Higher Education, 22*(7), 1520–1535. doi:10.1108/IJSHE-05-2020-0158

Acknowledgment

Dr. Mahmoud Bakkar:

First and foremost, I want to thank my amazing wife, Hanane. As much as it was important to the book's completion, she provided continuous courage and motivation. Thank you so much, dear.

In addition, I would like to thank Dr. Elspeth McKay for her continuous support of the book until it was published. I would also like to thank Holmes Institute Director Stephen Nagle for his unending support and trust. In addition, I would like to thank Professor Zijad Pita for his confidence and encouragement throughout my life. I also want to express my gratitude to my friend Reymond Voutier for his dedication and continuous support in sharing his concept of global trust. Finally, I am pleased to thank Professor Kevin Anthony Jones for his determination and motivation to support education through the teacher's global network. I thank him for his support of my book idea.

Additionally, I would like to thank the United Arab Emirates institutions for their unlimited support by awarding me the Golden Visa.

To conclude, I sincerely thank my family members, my brother Mohamed Bakkar and my two sisters (Zakia and Mona), for their constant support for my success throughout my life. And my brother's children (Yousef and Ghena) for their unlimited love.

Prof. Elspeth McKay:

Thanks to Dr. Mandi Axmann – Deloitte, Australia, and Dr. Marlina Mohamad – University Tun Hussein On Malaysia, who, through their expertise in special education technology learning design, shared their professional subject expertise, assisting the Editor in the manuscript peer-reviewing process. Their care and attention to detail were appreciated when there were conflicting chapter submission reviews.

Each chapter underwent a strict double-blind review process between authors to evaluate the submissions for this book. A special thank you is due to the generosity of some of these authors who agreed to review extra submissions – due to the promised anonymity, they cannot be singled out – all the same, they will know who they are.

Section 1
Industry 5.0: Optimizing Globalization and Entrepreneurship

Chapter 1
Global Trust Registry Plus (GTR +) Connecting G20 Vision With Actions in Industry 5.0 Era:
Global Trust Registry Initiative

Reymond Voutier
https://orcid.org/0009-0004-1149-0524
eNotus International Inc., USA

Mahmoud Numan Bakkar
https://orcid.org/0000-0003-4637-4035
Institute of Applied Technology, Abu Dhabi Vocational Education and Training Institute, UAE

Guillermo Pivetta
eNotus International Inc., Argentina

ABSTRACT

The chapter is based on the white paper distributed to the group of twenty (G20) and other global organizations concerning the global trust registry plus (GTR +) proposals for a Global Cities Social Venture Fund and affiliated commercial working capital services. This will assist recovery and generate new jobs following the Covid-19 Pandemic. The chapter explains the importance of trust measurement and proposes a global framework for trust calibration and evaluation. It recommends that the trust measure and the scoring register will help create more jobs by improving the efficiency and performance of all small and medium-sized enterprises. In addition, it will help organizations to collaborate and fill the gaps in globalization and communication between nations.

DOI: 10.4018/978-1-7998-8805-5.ch001

INTRODUCTION: ESSENTIAL PRIORITIES

The American sociobiologist mentioned that "The real problem of humanity is the following: we have Paleolithic emotions; medieval institutions; and god-like technology." — Edward O. Wilson, 1929 (Laszlo et al., 2020).

Rome, Italy – On July 5, 2021, an outreach event of the Business of 20 (B20) countries advocacy process, focusing on the main priorities identified through the work developed by the 132 members of the Task Force Finance & Infrastructure, made the recommendations below for conveying to the Group of 20 (G20) countries (B20 ITALY, 2021).

Two of the Four Priorities directly relate to the global trust registry plus (GTR +) concepts presented here.

- "Promote impact investing, sustainable finance, and financial inclusion by accelerating the adoption of global sustainability reporting and measurement standards and by fostering access by individuals and micro businesses to affordable financial products and services." (B20 ITALY, 2021)

"Support sustainable economic growth by fostering small and medium-sized enterprises (SMEs) access to capital, promoting open innovation ecosystems, accelerating digitalization and innovation processes in the financial sector, and increasing the efficiency of new global and regional value chains." (B20 ITALY, 2021)

These recommendations call for concrete and immediate action. Some officials and experts are calling for "the great reset" and what is being proposed here in this chapter for cities and jobs, is a transition phase of actions that must precede any great reset.

The G20 chose "people, planet and prosperity" as its theme for the Italian Presidency in 2021. Rebuilding economies and preparing for future shocks has been the subject of countless meetings, webinars, and high-level discussions. Making use of imagination and greater involvement by informed leadership, which is based on science, may result in better policies going forward, that will help citizens, governments, and businesses face the multiple crises affecting world economies. They can work together to try to recover and prepare a pathway to a better future for all (B20italy2021, 2021).

In that sense, the G20 Italian presidency has not only built successfully upon the Kingdom of Saudi Arabia's previous presidency agenda - but it has also set standards of excellence despite the difficulties encountered through the Covid Pandemic. As the G20 baton passes from Italy to Indonesia and then onto India, there is hope that governments are realizing that action is required.

As stated by Jesper Brodin, CEO of IKEA (Welch, 2022):

"Good intentions are important, words make them powerful, and in the end, it's only actions that create real change".

With this purpose, the Global Trust Registry Plus (GTR+) identified Cities and SMEs as having critical needs for new financial products and analytics, which must be at the heart of any economic recovery.

With new economic pressures, resulting from the pandemic, the information that follows provides effective plans for Cities, which shoulder a large part of the burden towards recovery and a sustainable future.

IMAGINATION, NEW INSTITUTIONS, AND CAPABILITIES

"Imagination is more important than knowledge. Knowledge is limited. (Goodreads, 2022)" – Albert Einstein

Thinking beyond tomorrow is becoming urgent and essential for continued innovation and progress, and this starts with people in communities and cities. Work and money values are changing rapidly. Short-term decision-making in societies has followed the way governments and enterprises have set their economic standards and goals. They are based upon built-in biases, with little concern for the planet or its people. The disillusionment of youth towards the government, business, and financial institutions is at an all-time high. There is a general feeling that things must be done differently.

Whatever reformulation of social and economic goals takes place in the decades ahead, it will need the cross-pollination of the widest and deepest inputs. It must include the citizens of today and the citizens of tomorrow. The speed of digital and neurological change will further accelerate. Democratizing our foresight will be part of democratizing money and the future of work. Digital technologies and currencies on the blockchain cannot work without this foresight.

Thus, many of these changes will happen in cities and towns. The strong voice of cities and their citizens will be required to bring this foresight of imagination and leadership to the life of a new economy. This can be achieved by forming new organizations, around a city coalition, that will involve the private sector and research institutions. This will encompass new financial know-how that can map economic recovery strategies over the transition.

These strategies will require incentives for the private sector, alongside investors and informed leadership. Together they will bring new thinking to the aspirations of projects such as the Sustainable Development Goals (SDGs) (United Nations, 2015).

Workable models need to be applied in cities and towns to plot pathways through the transition period to an inclusive and fair digital world. Advocacy, funding, and clear communications are going to be required to bridge the gaps in understanding that exist between institutions and the private sector. Today, complex global problems require innovative responses from institutions that do not currently exist. Massive inequality is now a universal problem. Without a new financial architecture and new goals for nations, cities, and their people, the already broken system is heading toward a total collapse.

Creating agile transitional structures is necessary for capitalism and democracy to continue and become better designed for dealing with the opportunities and challenges presented in the industry 5.0 digital world. There are three new entities suggested for formulating the global trust registry (GTR) platform in this chapter:

Each of these transitional structures offers the opportunity to make a positive

Figure 1. GTR entities

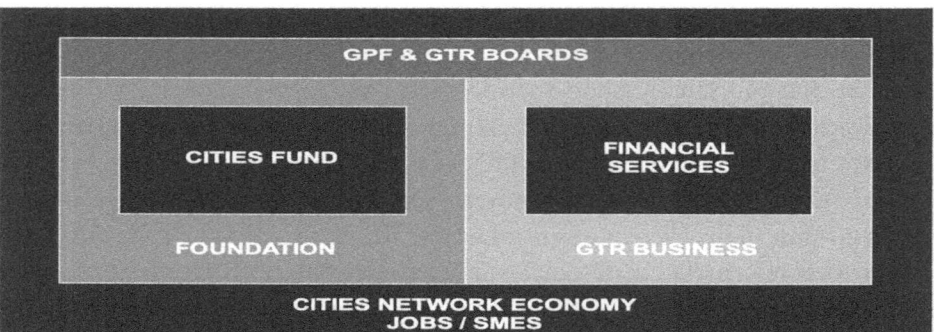

1. FOUNDATION + 2. CITIES FUND + 3. GTR BUSINESS

change in direction away from the set-in-stone perpetuation of outdated, and bad models, that haven't served societies well in recent decades. For instance, a vibrant and resilient workforce with meaningful jobs that provide citizens with dignity is going to be what measures a city's value, not its profits alone in the Industry 5.0 era.

The first two entities must have the necessary functional, yet independent roles that complement the initial GTR Business activities of providing working capital. By working in tandem, they better engage and motivate the private sector with an understanding of why supporting SDGs and other such goals makes sense for business (United Nations, 2015). Working together with a scalability model, cities that are at the center of this action can lead the way where international organizations have failed.

There is a need for a not-for-profit knowledge platform that helps build innovative partnerships in support of the 2030 Agenda. Its mission is to foster economic growth by convening thought leaders, policymakers, entrepreneurs, philanthropists, and investors to address global challenges working across the United Nations (UN) system (United Nations, 2015). It is also the co-founder of the Blockchain Commission for Sustainable Development, a multi-stakeholder framework that leverages blockchain and other frontier technologies (United Nations, 2015).

These innovative strengths provide the flexibility to guide the early stages of the proposal that bring technology and finance together to serve cities and their people, with meaningful jobs. The need is urgent and clear, and the GTR+ is an imagination startup, that will establish new thinking and relevant actions, building upon the largest network on planet Earth.

FOUNDATION, CITY FUND, AND GTR BUSINESS IN INDUSTRY 5.0 ERA

"Insufficient investment is due in part to inadequate infrastructure plans and an insufficient number of well-prepared investible projects, along with private sector incentives.

Structures that are not necessarily appropriate for investing in many long-term projects, and risk perceptions of investors." (United Nations - Addis Ababa Action Agenda, 2015)

The CITIES FUND envisaged will be a very new kind of facility to provide innovative city and business financing for a new economy. It will combine elements of venture capital with development finance to meet the needs of a digital world. It can therefore be able to undertake risks usually not found in debt funding models. This will be complimentary to Working Capital and Supply Chain Finance (SCF) products.

The situation for sustainable development including the infrastructure pointed out at the beginning of this chapter is critical because the pressure on governments, donors, and development banks has increasing limitations. Investors expect revenue streams to have a positive ROI, and those who see development impacts as important are the only small group on a much larger base that can be tapped through the city's scaling model. For this reason, the appeal of the Cities Fund needs to be aimed at this broader investor audience, including business enterprises themselves - the GTR model is designed to achieve this.

Research undertaken indicates that unless venture and equity-type investments can be incorporated with new fintech capabilities, there will be limited success in attracting private finance and institutional investment. Blended financial transactions have shown poor results. The world of money and transactions are changing at breakneck speed, and this will continue for many years.

CITIES FUND In the industry's 5.0 era, we must adopt innovative and hybrid models attractive to asset managers and understood by the business. It also needs to fit with Artificial Intelligence (AI) and Blockchain innovations as Fintech entities challenge the banking system.

The Sustainable Development Goals and the Addis Ababa Action Agenda, with a vision of "billions to trillions" are unlikely to materialize by relying on governments and donors. Yet it is certain that cities and their business enterprises must find a way to survive, and once again flourish. This is true for so many other areas, such as education and entrepreneurship within cities, all requiring a new and better approach to funding and a better data environment.

During discussions with fund managers and organizations such as Business for Inclusive Growth (B4IG) formed after the Paris G7 in 2019, there are clear signs that there is a potential to establish this new kind of City Fund that can become as attractive as a venture capital or other type of digital investments.

Building the Right Team

The Cities Fund will attract excellent management talent and a new breed of technically capable young executives. Many banking and fintech executives with a desire to focus their work on impact investment are already leaders, familiar with changing technologies. This will bring talent to the undertaking and allow the Cities Fund and GTR working capital syndicate to consider investments with a long view to achieving maximum performance without fear of defaults. While the GTR Business working capital finance syndicate is a commercial and private operation, it can work closely with the Cities Fund, complimenting it and setting in motion the big ideas for a new and fairer economy. The rapid rise of digital and crypto activities makes this essential.

Actions shape the future, and new awareness is growing throughout businesses worldwide. In many cases, those in business are far better informed than governments, realizing that without new and fairer social contracts, markets will be destroyed, and without markets, the business has no purpose and society is at risk.

WHY CITIES NOT NATIONS

GTR Business Working Capital and Analytics

These activities will be the basis for launching the GTR program in cities to commence helping business enterprises and providing the framework for the trust services and people analytics that will follow, including the Cities Fund.

"Let cities, the most networked and interconnected of our political associations, defined above all by collaboration and pragmatism, by creativity and multi-culture, do what states cannot. Let the mayors rule the world."(Barber, 2013)

During 2021, under the leadership of the co-chair cities of Rome and Milan, the Mayors and Governors gathered as the Urban 20 (U20) to call on G20 leaders to partner with cities in achieving human-centered, equitable, carbon-neutral, climate-safe, inclusive, and prosperous societies. In that sense, the proposals leading toward prosperity were:

1. **Adapting to the future of work and a just transition** – The G20 should empower cities to address the structural changes, including labor polarization due to the digital economy, that are reshaping the future of the work field (U20 Rome-Milan, 2021). They urged investing in decent and well-paying jobs that generate revenue from climate mitigation, supporting a just transition for fossil fuel workers, and developing national strategies that account for local jobs' impact and increase women's participation in the labor force. G20 should provide equal access to quality education for all, employment opportunities for youth and people with disabilities, professional and vocational training, and upskilling in soft and digital skills (U20 Rome-Milan, 2021). In summary, they must tackle the administrative, economic, and technological barriers that impact our communities.
2. **Strengthening local democracy** – They called on the G20 to protect and elevate local decision-making and public participation mechanisms, both in physical and digital channels, by developing legal and institutional frameworks that fully include and respond to the needs of cities, including data, transparency, and freedoms of press and expression (U20 Rome-Milan, 2021).
3. **Promoting fiscal autonomy** – They called on the G20 to rebuild cities' fiscal autonomy to secure revenue streams for better planning, borrowing, and investment and establish enabling conditions to create an ecosystem of public and private financial partners that can mutually support each other and aid in securing much-needed financing (U20 Rome-Milan, 2021). This included

enhancing the capacity of cities to guarantee compliance and integrity in public procurement and cultivate new sources of revenue, for example, through data production and circular economy and decarbonization models. They should scale up blended financing models, backstop innovative means of multilateral and cross-border finance, reinforce the capacity for local finance, and act to increase flows of domestic and international capital for transformative financeable investments, especially in underserved markets while ensuring that such investments do not harm local livelihoods (U20 Rome-Milan, 2021).

4. **Fostering local economic development** – They called on the G20 to foster production and consumption models based on proximity, sustainable tourism, and green manufacturing, and to support local micro, SMEs as the backbone of our economies and connect them to international value chains (U20 Rome-Milan, 2021). They must continue to promote and harness inclusive innovation and multi-stakeholder engagement and collaboration through social platforms. They must recognize the importance of cities as part of their local economies and help maintain the consistency and sustainability of local public services when economic shocks impact local budgets (U20 Rome-Milan, 2021). Furthermore, they must harmonize and expand international standards and certification mechanisms for companies and investment funds to contribute to the SDGs more directly at the local level (U20 Rome-Milan, 2021).

5. **Protecting digital rights** – They called on the G20 to acknowledge the lasting impacts of digitalization on equity and citizen engagement during and after the pandemic by urgently enabling new forms of data regulation and digital governance, which adapt human rights to the digital age by fostering digital rights, transparency, and privacy, and which consider the regulation and tax consequences of operating and providing digital services internationally (U20 Rome-Milan, 2021). They should support efforts to bridge the digital divide (connectivity, devices, and skills) and ensure that future investments in smart cities follow open and ethical digital standards, make technology and data affordable, citizen-controlled, and universally accessible, support digital literacy, and guarantee freedom from censorship and discriminatory bias within algorithms and resulting artificial intelligence systems (U20 Rome-Milan, 2021).

Roles for Cities

Promoter - Be a role model by implementing exemplary practices. Identify circular opportunities for industrial symbiosis and potential collaborations across usually unrelated sectors. Develop a strategy for a circular economy with clear goals and targets and include employment opportunities and associated skills.

Facilitator - Establish horizontal and vertical coordination mechanisms to align policies at various levels of government and across city departments. Connect with the business community: support knowledge and practice sharing. Connect with universities and citizens to provide solutions to the city's problems. Facilitate the connections between urban and rural areas for the food sector, construction, biomass, and delivery sectors.

Enabler - Identify gaps and ways forward to support the adaptation of laws and regulations for circular business to happen. Establish green public procurement containing circular criteria. Stimulate entrepreneurship. Generate an information system for the circular economy.

BRINGING CITIES TOGETHER

"If cities are to play the major role they are assigned in the fields of economic development and struggle against climate change, their funding capacities need to be rapidly and massively increased." (U20 Rome-Milan, 2021)

The need to develop a GTR system is crucial to strongly focus on generating a trusted movement - one that would benefit cities and their SMEs with new analytics and supply chain finance products. This remains relevant as economies need to be rebuilt and transitioned to the new economy.

GTR will enable industry leaders with the mission of democratizing access to capital markets. To enable Financial Institutions, Corporates, and their Supply Chains to drive growth by transforming access to working capital within cities.

Banking has become hyper-competitive and less profitable, and in almost every city around the world, customers are being poached by non-banking entrants, leading to a challenging banking environment. Banks, however, remain key channels in most cities to reach smaller enterprises and can provide part of the answer to a new and more effective ecosystem that supports these enterprises. Applications being developed based on GTR concepts with financial syndicates can improve profitability, allowing banks to address competition head-on. These same applications are well suited to help banks in their market positioning in cities around the world.

Advanced Trust Scoring, a technology platform that will enable banks to not only retain but grow their customer base, is the goal of the GTR platform concepts to be applied within cities. Services can be customized for banking partners regardless of size, helping the capital needs of the bank's customers. Some banks may be unable to offer certain services due to regulatory constraints such as jurisdiction, concentration risk, Basel III, or Consolidated Know Your Customer (KYC) risk management regulations. These are trust-related issues that are plaguing much of the crypto market and players in the Fintech field.

Reaching SMEs Through City Partnerships

Large numbers of SMEs can be reached through the GTR city strategies with the cooperation of the cities and channel partners, and over five or more years, this will also lend itself to a mutual style of fund involving network members, so they can benefit directly from financial and insurance products. The power of the GTR strategy with Cities is that it creates synergies between key players. SMEs form the backbone of most economies around the world. Representing one-fifth of global banking revenues. SMEs generate around $850 billion of annual revenue for banks - a pool expected to grow annually over the next decade (Barua et al., 2019).

Figure 2. GTR Reaching out to SMEs through city partnerships

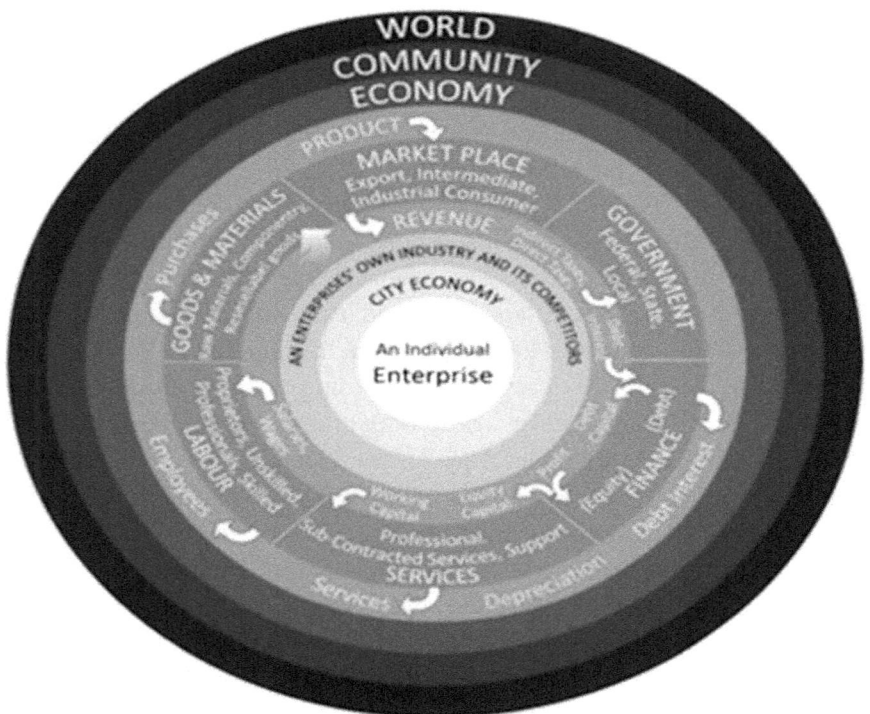

Winning the SME Market Requires Banks

Finding a balance between customer service experience and the cost-to-provider of such services to smaller enterprises has been difficult. This has resulted, in many banks not having prioritized the needs of SMEs despite the vast potential they offer – and many of these smaller businesses still believe banks ignore their needs (World Trade Organization, 2016).

New customer propositions and better service models enabled by technology are creating new opportunities for more lucrative returns. FinTechs are entering the business, at pace, as are the "big tech" firms, with innovative service models that reduce costs and increase revenue. Their offerings include traditional banking products, often packaged with other business services like payroll, invoicing, or tax preparation. Banks can reinvent themselves and regain market share by using little imagination.

Banks are in a very favorable position to build SME ecosystems because of their trust-based customer relationships. The data they have through these relationships and their ability to invest in services that help business owners provide them with a clear advantage. This is where analytics adds value to banks, customers, and cities.

There are several key reasons why the concept of GTR - SMEs residing in cities - where GTR will promote in conjunction with the city administration - should find the idea of being part of a larger global community attractive, especially due to business and financial services such as working capital.

Access to a more democratic form of financing and new working capital products is essential for most enterprises, and these services will be introduced through banking and fintech channel partners. As part of the city's future of work initiatives, there will be other more beneficial reasons for SMEs to participate in GTR membership.

Membership presents a great opportunity through a trust score to obtain other future benefits. The creation of the Cities Fund is one of these, as the largest of the problems cited by SMEs worldwide is access to capital.

CITIES FUND is being developed to complement the offering of GTR concept syndicate working capital products, which is seen to establish a massive fund for small enterprises, so they can gain the financial benefits enjoyed by larger corporations. The development of this mutualization model will provide digital tokenization of member ownership assets; for the first time, allowing well-managed smaller enterprises to raise capital as easily as larger corporates. It is envisaged that, over time, as much as 30% of the Cities Fund could result from enterprises themselves creating a powerful market mechanism.

INVESTORS CAN RAPIDLY INCREASE EMPLOYABILITY

"Purpose is not the sole pursuit of profits, but the animating force for achieving them. Profits are in no way inconsistent with purpose — in fact, profits and purpose are inextricably linked." (Fink, 2019)

Cooperation among investors is the key to implementing new ideas, and new opportunities are being presented every day. More than a third of the world's invested capital — estimated at $19 trillion— is controlled by the world's hundred largest asset owners. Nearly two-thirds of this money is in pension funds, while the remaining third is in sovereign wealth funds (pionline.com, 2018).

The fifteen largest asset managers collectively handle nearly half of the world's invested capital (Henderson, 2021). They include BlackRock, which currently manages just under $7.0 trillion; the Vanguard Group, which controls $4.5 trillion; and State Street, which has $2.5 trillion under management (Henderson, 2021). These asset managers need to be shown opportunities that can work for them and their investors that will benefit society.

GTR is Servicing as a Global Platform for Analytics With the Right Information

The foundation of the Global Trust Registry's platform is people. It is the variability of people, that has compromised many well-meaning programs in the past.

The Global Platform for People Analytics is bringing proven science and AI-assisted analytics to ensure that the GTR's applications are robust and data-centered, bringing the right resources in the right way to create sustainable change and city value.

The GTR concept includes a dynamic and powerful Global Centre for People Analytics. The Centre's unique ability to capture workforce DATA and translate it into practical problem-solving for employability is a powerful component of the GTR's strategy. This DATA brings great differentiation to the GTR's initiative for cities, SMEs, and investors.

Science has recognized that people have hard-wired personality traits and cognitive abilities for decades. This is foundational for job success and is the keystone for anything involving behavior or learning. These can now be easily measured, but the innovation that makes these game-changing transformations possible is the simplification of the information revealed in the DATA.

Executives, educators, and government leaders will find the GTR to be an invaluable resource for guiding them in applying behavioral DATA to their planning and operational strategies. AI experts should be able to correlate each City's workforce DATA with other economic and social data to create and optimize their resources.

Figure 3. GTR servicing as a global platform for analytics

For the Cities: Resilience and Employability

For cities recovering from the COVID-19 pandemic and its catastrophic impact on worldwide labor markets, employability has become paramount. How does one make themselves Employable? Old job titles may have little connection to the new jobs demanded by the redistribution of the workforce on businesses. At the very least, years of experience in one job in the old normal has a different, and usually lesser, value in the new normal. For the cities' labor markets, Resilience is the critical objective. How does the labor market create viable connections for jobseekers amid a storm of change? Even in a more stable business economy, there were few success stories in the standard factory closing and re-skilling programs. Workforce training initiatives were forced to admit that billions had been spent training jobseekers for jobs they did not hold a few months later. With new job titles and an evolving assortment of working conditions, the path to anything approaching Resilience is unclear. The extraordinary thing is that both Resilience and Employability can be achieved at the same time by combining proven science and technology with remarkably simple digital applications.

Every job in the world depends upon a simple set of critical job behaviors. Hundreds of jobs have been translated into these job behavior metrics. The wizard system has been designed to enable anyone to do any type of job. This is the foundation of a Resilient labor market, every job is defined and described by a common set of metrics that is the same set of metrics describing the job strengths and abilities of jobseekers.

Activating this Resilient labor market requires employable jobseekers. It requires a completely new and more specific definition of Employability, one that links the true value of the jobseeker with the needs of the labor market. In an economically dynamic labor market, each person would know their Employability.

The traditional definition of employability has been: *"Employability is having a set of skills, knowledge, and personal attributes that make a person more likely to gain employment and be successful in their chosen occupations."*(QualityRese archInternational, 2022) As employment for so many people have become more transient, the value of a particular set of skills or knowledge has become less certain. Millions of IT workers have awakened one day to discover that the software they have mastered has been made obsolete by a technological breakthrough. Similarly, workers manufacturing quality products in a long-established factory have been faced with a world in which those products are no longer needed. The skills and knowledge that have value in one job do not necessarily maintain that same value in another job. Even when the value is there, it is rarely recognized by the next employer. The key to Employability is simply having the ability to perform the job in question successfully. The availability of jobs is always changing, and COVID-19 has devastated numerous job categories.

Every job depends upon the performance of a set of individual job behaviors, and all job behaviors depend upon hard-wired personality traits and cognitive abilities. These are the strengths that are in each person. They are always there. They do not go away or change appreciably. When the strengths of an individual match the strengths needed for a certain job, the Employability of that person is evident. These behavioral metrics present a common language that transforms the concepts of a Resilient labor market and the Employability of jobseekers into a practical and operational reality.

For the SMEs: Productivity and Sustainability

SMEs are a powerful economic force around the world, as indicated throughout this chapter. They usually form the greatest percentage of businesses within every country; the greatest percentage of employment; and a significant percentage of the GDP (World Trade Organization, 2016).

Unfortunately, SMEs are also subject to a disappointing failure rate. Those failures are not caused by government regulations, and few are the result of economic circumstances. Most SME failures happen because the salesperson could not sell the product or service, the manager could not manage the business effectively, or the production manager could not recognize quality. Essentially, businesses did not have the right people for the right jobs. IBM can hire a hundred poor performers and there is little effect on the whole enterprise. If a small business makes one bad hire, or if the founder or one of the key people is a bad fit for their role, that business is unlikely to survive. It will never flourish.

The GTR global platform for people analytics should provide a simple tool that enables SMEs to ensure that they have the right people for the right jobs.

Research has proven that using behavioral DATA to match people's strengths to those necessary for the job dramatically improves productivity by as much as 25%. Introducing that capability to the SMEs within any city is a game changer for that economy. The GTR platform educates city and business leaders and entrepreneurs, making the impact sustainable.

For the Investors: Assurance and Confidence

The success of any business investment depends upon the people within the business. That is the variable that creates the risk. Using behavioral DATA, the GTR global platform for people analytics can inventory the operational strengths of the executive team and the workforce of any company. This DATA enables GTR applications to provide a level of assurance to the viability of the enterprise and the operational capability of its people. That ability to bring a measure of confidence to investors is unique to the GTR's financial programs.

The GTR Concepts and its Platform for People Analytics

By integrating advanced behavioral DATA, AI-assisted analytics, and extensive financial resources, GTR presents an array of advantages for Cities, SMEs, and investors unique to the world. It can be transformational for all of these.

GTR PLATFORM SERVICES ECOSYSTEMS

Why do GTR Platform Services Ecosystems and People Analytics Offer Cities & SMEs a Winning Edge?

When does it make sense to invest in human capital? One of the reasons why GTR places such importance on the Trust is that people analytics for Cities are at the center of all decision-making.

SMEs are not only the largest business market, but they also have significant financial needs. So, it's easy to understand the attraction of ecosystems such as Alibaba, Amazon, or Salesforce offer. The ecosystem offers a broad range of services that SMEs must have, including but not limited to banking, through a single integrated platform.

This provides the service convenience that SMEs seek and reduces the amount of time they spend searching for and accessing services. An ecosystem can also make new and better services possible – think AI and Blockchain.

Companies that offer these kinds of solutions will stand to gain market share. By generating new revenue streams, increasing customer satisfaction, and easing customer operations through data that can be used in multiple ways.

While some other players — such as technology giants and telco companies — share some of these characteristics, to a degree, it is transactional data that truly makes a difference. Banks have access to a vast amount of transactional data that sheds light on many aspects of a company's owners, management, behavior, and activities. This data can provide a profound advantage in the race to offer ecosystem services, and GTR is proving to be the best DATA model.

The Complexity and Dynamism of Cities Data Ecosystems behind a Territorial Approach (Cooperation with Global Research Leaders - OECD Regions and Towns)

The complexity and dynamism of Cities as Systems of Systems can be represented by many Factors and Elements that are mediated by the interactions between Government, Business, and Innovation Networks. Thus, generating behavioral and micro-economic data to be collected and analyzed to match the GTR platform concepts of services.

A big challenge of today is the business case for meeting the UN's Sustainable Development Goals to save the world, with a cost of $12 trillion in operation (United Nations, 2015). The funds required will be difficult to raise at best, but without close and workable cooperation with large sections of the private sector, the task is impossible. The territorial Approach to SDGs is an effort by the Cities, Urban

Policies, and SDGs Division of the Centre of Entrepreneurship, SMEs, Regions, and Cities at the OECD.

Figure 4. Cities data ecosystems

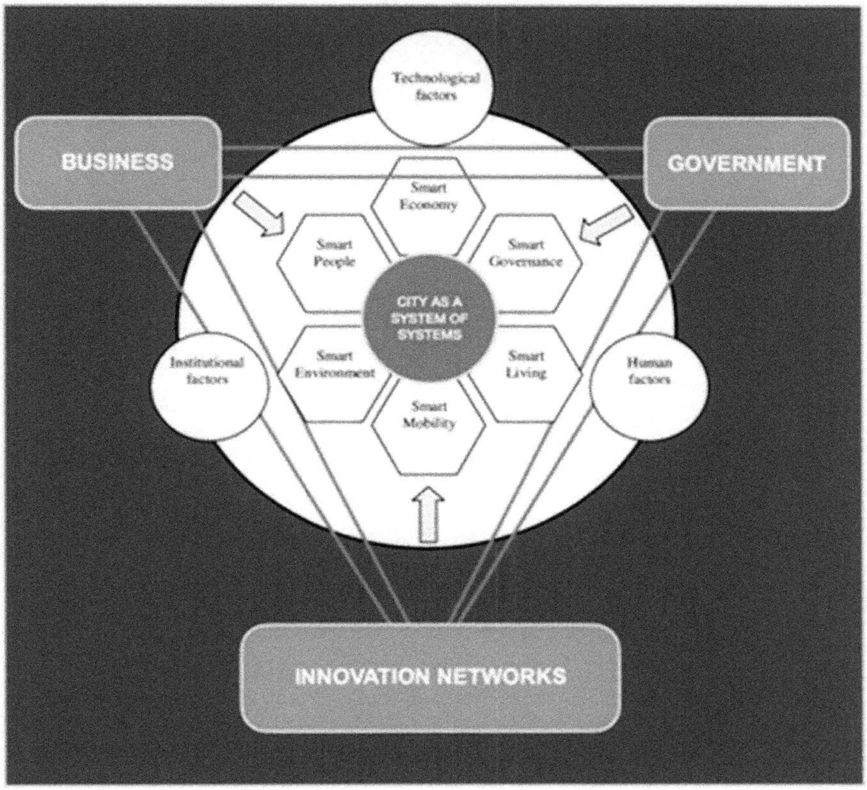

CONCLUSION: BRINGING IT TOGETHER

Cities are a perfect example of a System of Systems - they provide a scalable model for GTR to build a new kind of digital sister cities network, one that will assist economies in transition to a digital future. Global and local services essential for job creation will merge on a common services platform supported by finance and analytics necessary for the new economy.

Data will be fed back through an AI-powered trust system to allow the Centre for Peoples Analytics to provide essential insights for Government, Business/SMEs, Innovation networks, and of course Investors.

Figure 5. Bringing it together

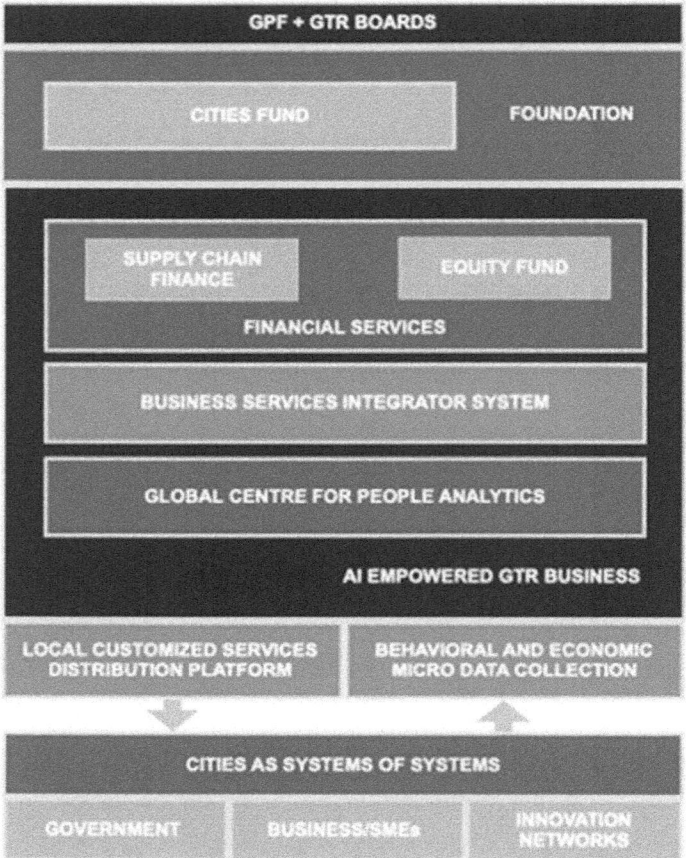

While the GTR platform should be designed to better connect enterprises with their peers and counterparts and ensure access to practical support such as working capital - it is the mission of the Cities Fund to provide long-term assistance and support to cities for transitional needs.

The three entities of the GTR initiative (GTR +) bring together the new engine essential for helping cities to help their business enterprises. A bold and different approach remains practical to assist economic recovery and better prepare cities for the digital transition and a new economic era.

There has been much discussion during the pandemic about the big issues and thousands of recommendations made about how to deal with potential future health crises, the economy, and the strategies to deal with climate disasters. All can perhaps be assisted with technology and better policy foresight - however, only action on connected issues can keep the economic machinery going during the future of work transition to the new economic era. Blockchain, digital currencies, and tokens alone are not the solution.

The basic issues of poverty, economic imbalances, and a living wage are all priorities, as are inclusivity and sustainability. Fortunately, there are very hopeful signs from business leaders across the globe that there is a change in the bigger purpose. This change is a deeply rooted understanding of the wisdom among our youth, who will need to take leadership positions that will determine what kind of future they and their children have.

To create new social contracts for an age that has seen the abuse of social media and witnessed the continual destruction of our environment and many other norms that benefitted the generations post World War 2 (WW2). The consensus ideas and institutions that emerged during this period are fast approaching the use-by date.

The old institutions may not be fit for purpose, and newer ones are not yet fully designed or ready for the future. Going into space will not solve the pressing problems here on Earth in the short term, and perhaps never. Even the most optimistic talk at the big institutional tables about projects such as SDGs will require all the help they can get.

The lights must be kept on, and the dignity of workers and small enterprises not only be preserved but improved. This must be achieved while brainpower and creativity are fostered to find new economic models and purposes.

REFERENCES

Barber, B. R. (2013). *If mayors ruled the world: Dysfunctional nations, rising cities.* Yale University Press.

Barua, C., Gati, B., Lajumoke, T., Taraporevala, Z., Havas, A., & Radnai, M. (2019). Beyond banking: How banks can use ecosystems to win in the SME market. McKinsey & Company, 1–13.

Fink, L. (2019). *Purpose & profit.* Harvard University. https://corpgov.law.harvard.edu/2019/01/23/purpose-profit/

Goodreads. (2022). *Quotable Quote.* Goodreads. https://www.goodreads.com/quotes/556030-imagination-is-more-important-than-knowledge-for-knowledge-is-limited

Henderson, R. (2021). *Reimagining Capitalism in a World on Fire: Shortlisted for the FT & McKinsey Business Book of the Year Award 2020.* Penguin UK.

B20italy2021. (2021). *ABOUT B20*. B20. https://www.b20italy2021.org/b20/

B20ITALY. (2021). *Finance & Infrastructure, Policy Paper 2021*. B20 ITALY. https://www.b20italy2021.org/wp-content/uploads/2021/10/B20_FinanceInfrastructure.pdf

Laszlo, C., Cooperrider, D., & Fry, R. (2020). Global challenges as opportunity to transform business for good. *Sustainability (Basel)*, *12*(19), 8053. doi:10.3390u12198053

Pionline.com. (2018, November 12). Pension funds dominate largest asset owners. *Pion Line*. https://www.pionline.com/article/20181112/INTERACTIVE/181119971/pension-funds-dominate-largest-asset-owners

QualityResearchInternational. (2022). Employability. In *Analytic Quality Glossary*. https://www.qualityresearchinternational.com/glossary/employability.htm#:~:text=Employability%20is%20the%20acquisition%20of,whether%20paid%20employment%20or%20not).&text=Employability%20usually%20refers%20to%20the,but%20this%20includes%20self%2Demployment.

U20 Rome-Milan. (2021, June 17). *Urban 20 calls on G20 to empower cities to ensure a green and just recovery*. Urban 20. https://www.urban20.org/wp-content/uploads/2021/06/U20-2021-Communique-Final.pdf

United Nations. (2015). *The 17 Goals*. UN. https://sdgs.un.org/goals

United Nations - Addis Ababa Action Agenda. (2015). *Addis Ababa Action Agenda of the Third International Conference on Financing for Development*. UN. https://www.un.org/esa/ffd/wp-content/uploads/2015/08/AAAA_Outcome.pdf

Welch, N. (2022). *Is humanity at the centre of your leadership and team culture?* Natalie Welch. https://www.nataliewelch.com.au/is-humanity-at-the-centre-of-your-leadership-and-team-culture/

World Trade Organization. (2016). *World Trade Report 2016: Levelling the trading field for SMEs*. WTO. https://www.wto.org/english/res_e/booksp_e/world_trade_report16_e.pdf

Chapter 2
Globalization and Entrepreneurship in the Industry 5.0 Era

Mohammad Izzuddin Mohammed Jamil
Universiti Brunei Darussalam, Brunei

ABSTRACT

The definition and concepts behind globalization are sometimes confusing for the masses, and studies on globalization often focus on the macroeconomic level. Globalization is the interdependence of people and businesses everywhere that leads to global cultural, political, and economic integration. The purpose of this chapter is to provide background and insight on globalization, from concepts to linking globalization to micro, small, and medium-sized enterprises (MSMEs). Also, to showcase how favorable globalization conditions are beneficial for the growth of MSMEs, which can ultimately result in them expanding their area of operations in terms of sales, number of employees, and size. This chapter presents George Yip's broader framework of globalization that highlights the conditions needed for MSMEs to grow and flourish, including external factors. A framework is proposed to highlight the summarized framework of the external factors and conditions of globalization that influence the growth of MSMEs.

INTRODUCTION

Globalization is a collection of hazy and disparate notions from several areas and semi-scientific conceptions. However, from a geographic point of view, the world of startups and entrepreneurship, along with many large corporations in various

DOI: 10.4018/978-1-7998-8805-5.ch002

industries, is in profound transition. Since the 20th century, the trend has shifted from a series of discrete national initiatives that only rely on selling local exports and domestic products at the regional or national level to a more integrated global industry whose economies are increasingly interdependent on one another (Ritzer, 2008). In the case of startups and formal Micro, Small, and Medium-sized Enterprises (MSMEs), the business ties are increasingly global, which are often accompanied by strong regional patterns at the operational level; managers are slowly beginning to realize the importance of working together and integrating with a globalized market economy.

Market saturation, political pressures, dynamically changing business environment, unexpected economic recessions, and rapid advancement of technology have encouraged certain startups and MSME owners to muster the courage to not only sell their products at local scope but to sell them at the international level such as by exporting them overseas. Many startups are content to run their small businesses and abstain from expanding due to high-risk aversion. However, this comes at the risk of failing to adapt to a rapidly changing business environment. Not accepting one's business as part of a larger network of businesses runs the risk of being outdone by competitors or losing consumers worldwide due to not being flexible to adapt to changes (Watson, 2003).

As globalization accelerates, more enterprises join the market. Because of the previously mentioned accessibility, small enterprises face significantly increased rivalry worldwide. There will certainly be ten other firms providing comparable items to the one MSMEs are attempting to market. More enterprises enter the market as the globe gets more globalized, and small companies worldwide face increased competition due to the previously mentioned accessibility from globalization.

Thus, MSMEs must have the right mindset in this era of globalization. MSMEs are increasingly more relevant in the face of globalization, which has become an influential force in global trade. However, MSMEs in both developed and emerging economies face many common challenges in the globalization era, including the absence of adequate and timely finances, limited capital, and access to the international market and knowledge (Anand, 2015).

Globalization

Globalization is one of the most common terms used to describe political, economic, social, technological, environmental, and legal changes to an economy due to integration and interdependency among people and businesses from across the globe (Chew, 1993; Yameogo et al., 2021). On a more official note (Ch et al., 2011), the International Monetary Fund (IMF) adequately defines globalization as the rise of economic interdependency among economies worldwide. This is done with an increasing volume and variety of cross-border transactions in goods and services, free international capital laws, and more rapid and widespread diffusion of technology (Heimberger, 2020).

In this era of globalization, academics, practitioners, managers, and consultants have all recognized that studying industries, strategies, and organizations in a global context needs to be regarded as the norm. This is especially the case for small businesses such as startups and MSMEs. Moreover, the discourse has been made that success or failure of an organization in the twenty-first century will depend on whether it can compete effectively in world markets (Hax, 1989).

While various types of globalization fields (Dreher et al., 2008) can be applied to today's context, such as financial, political, cultural, and information, industrial globalization is arguably the most crucial. Industrial globalization pertains to the rise and expansion of businesses beyond the local and national level to go overseas (Narula & Dunning, 2000), and organizations that are heavily engaged in international trade are known as Multinational enterprises (MNEs), multinational corporations (MNCs) or transnational companies (TNCs).

Thus, for this chapter, globalization is defined from a business or organizational point of view, whereby both small and large enterprises have transactions that cross international borders, whereby business transactions are conducted between economies. The enterprises effectively term themselves international businesses. The growth of globalization has been driven by several factors, such as advances in transportation and telecommunications (Crafts, 2004). This is further emphasized by the rapid growth of the Internet, which has facilitated the global sharing of information and resources (Block, 2004), besides providing an efficient channel for the direct distribution of certain goods and information services. In addition, the spread of liberalization has resulted in removing trade barriers and restrictions, causing investment activities to increase.

Using the KOF Globalization Index indicator, it is possible to measure an economy's globalization level (Vujakovic, 2009). The KOF Index of Globalization covers the dimensions of economic globalization, social globalization, and political globalization. Globalization is rapidly advancing, especially in developing or emerging economies such as in ASEAN + 3 (Gygli et al., 2019), where it can be noticed that all ASEAN+3 countries show a substantial increase in the KOF index over the period 1970-2012. In comparison to developed economies, such as the G7 countries, it is noteworthy that globalization has been decreasing or fluctuating (Gygli et al., 2019), resulting in them achieving lower GDP gains from globalization.

Another factor that influences the growth of globalization is the international movement of commodities such as money and information and the movement of people or human resources. Social networks, too, may be an important stimulus for growth and internationalization, influencing the entrepreneur's ability to take advantage of market opportunities and external resources (Hoang & Antoncic, 2003). First, however, an in-depth study of the conditions suitable for globalization's growth potential is needed at the industry level.

Another purpose of this chapter is to provide an overview of globalization, its concepts and drivers, and how stages of economic development each have different pillars or characteristics for MSMEs to grow. Studying globalization is sometimes controversial, as debates rage on whether globalization brings about global benefits or drawbacks. Justifications for synthesizing theories on globalization and the external factors influencing the growth of MSMEs are discussed. It is vital to examine the globalization drivers in an industry to determine the favorable conditions for an MSME to grow and set up an International Business in a process known as internationalization. A proposed theoretical framework is also shown in this chapter to showcase the newly synthesized theories. This chapter also suggests internalization strategies, as the survival of MSMEs hinges on making informed decisions on which strategic options to choose.

Debates on Globalization

The controversial nature of globalization entails that some sects favor globalization, while other denominations are considered anti-globalization (David & Barwinska-Malajowicz, 2018). Globalization is blamed for many socioeconomic shortcomings. The most obvious defects come from an environmental perspective, where globalization has resulted in environmental problems (White, 2010). One prominent example is globalization's influence in transforming Thailand's economy into an export-oriented center; many dams were built in the 20th century, resulting in contaminated rivers and inhabitable habitats for plant lifeforms (Molle, 2007; Tonts & Siddique, 2011).

Globalization brought about a conventional manufacturing strategy focused on a linear economy (Máliková, 2020), which meant that items were not created to be reusable, repairable, or even recycled afterwards. Following use, the consumer immediately regarded these goods as garbage and spent the rest of their lives in landfills (O'Bannon et al., 2014). As a result, the quantity of waste created every year in the globe has expanded dramatically due to globalization and industrial industry, consequently damaging the environment and resulting in the economic system's linearity reaching its limit.

Consumers purchase an increasing number of things that will expire after a certain period or lose their worth for money; eventually, they will not be consumed in time and will be discarded. Furthermore, the mountains of rubbish are left as an unwelcome legacy for future generations. However, not all trash is truly waste. More than half of this trash has the potential to be utilized as a secondary raw resource. This fact is also acknowledged by businesses that, to safeguard the environment, they use the circular economy idea (R. Stahel & MacArthur, 2019). Fortunately, the circular economy is an economic strategy that aims to eliminate waste and continuously use resources. Circular systems make use of reuse, sharing, mending, and recycling. This activity means that after usage, all items and materials will be reconnected to their cycle, where they will become sources for manufacturing new products and services, which can counteract the effects of globalization.

From a social standpoint, globalization has introduced Western culture in different countries, producing bicultural identities and self-selected cultures (E. Lee & Vivarelli, 2006). The integration of the two cultures has created identity confusion and challenges to adapt to either. To an extent, economists have argued that the financial crisis of 2007 and 2008 is partly due to globalization (Asgary & Walle, 2002). This activity is because globalization has caused the financial crisis, which originated in the United States of America, to reach a global level. Some argue that globalization still needs to achieve its objectives of economic equality and improving the social standard of living (Velasquez, 2000). Due to the global integration of some groups happening alongside the marginalization or exclusion of others, globalization can be described as an "uneven process", such as in Africa (Loretta et al., 2015). The recent outbreak of the COVID-19 pandemic further serves to justify anti-globalization. It was argued, due to the economic and cultural integration of societies into a single market entity, globalization paved the way for the virus to easily spread to other areas (Farzanegan et al., 2021); most commonly via aviation, resulting in the said industry being shut down and badly affected by the pandemic. Indeed, income inequality has worsened since the (Coe, 2007; Tridico & Paternesi Meloni, 2018)

On a more positive note, researchers have argued that globalization has had many positive impacts. In theory, the objectives of globalization include the promotion of free trade (Goldstein, 2010) to ensure reduced barriers such as tariffs, import duties, and quotas, the creation of job opportunities around the world, especially with the establishment of multinational corporations (MNCs) due to MSMEs becoming international businesses, and economic growth from due to economies of scale (Urata, 2002), technological innovation and foreign direct investment (FDI). In practice, however, globalization has broadened access to goods and services, making them accessible for people regardless of socioeconomic background. This activity is especially useful for countries that do not have the required resources or are incapable of producing certain goods and services, known as the absolute advantage (Goldstein, 2010).

One of the most important benefits of globalization is the transfer of technology globally (Mayer, 2002), especially sharing the latest technologies with developing countries. Technology transfer is an important part of the technological innovation process. It involves promoting scientific and technological research and the associated skills and procedures to wider society and the marketplace, ensuring everyone has equal access to technology and preventing the digital divide (Jamil & Almunawar, 2021).

Globalization has also strengthened cultural awareness worldwide (Sifakis & Sougari, 2003). A globalized society boosts the rate at which people are exposed to the culture, attitudes, and values of people in other countries. This exposure can inspire artists, strengthen ties between nations, and dampen xenophobia.

Arguably, the most crucial reason for globalization is enabling a nation to mitigate economic risks by sharing them or diversifying them with others (Norman, 2011). In other words, the world will absorb the economic shock of one nation, as evidenced by the financial crisis of 2007-2008, where the consequences of the financial crisis are shared and felt by not just the United States but countries in Europe, Africa, and Asia. The Great Depression in 1930, another severe economic depression that started in the United States, was spread worldwide due to globalization (James & James, 2009). Hence, the conclusion derived from both sides is to find an idea that benefits everyone or a synthesized framework that showcases conditions that can help small businesses grow. The framework proposed in this chapter, rather than focusing on either side of globalization; instead showcases the requirements that an economy that is a part of the globalized must fulfill to promote the growth of businesses, particularly small and medium-sized enterprises.

Chapter Focus

Research Gaps

Globalization is not discussed here in the context of entrepreneurial studies. Instead, this chapter's contribution is to promote studies on entrepreneurial globalization. Globalization and entrepreneurship are intertwined; for starters, globalization promotes technological entrepreneurship by encouraging the growth of innovative ecosystems. This issue might involve collaboration between startup companies and major international corporations. Firms have been able to specialize due to globalization and enhance the intensity of R&D, innovation, and capital in their production. Globalization has made it simpler for new businesses to compete with established ones. Finally, the trade industry has grown the number of people it employs through exports and imports. Thus, more globalization studies in the context of MSMEs are needed.

Furthermore, the George Yip model is limited in globalization literature, prompting this chapter to showcase its benefits. According to Yip, *"the global corporation does not have to be everywhere but has the potential of going anywhere, deploying any assets, accessing any resources, and maximizing profit worldwide"* (Yip George S.,

2004, p. 2004). This statement is useful for MSMEs aiming to succeed internationally by considering the drivers of George Yip's Model of Drivers framework.

George Yip Model of Drivers Framework

In his book, The Total Global Strategy, George Yip (1992) introduced one of the most important theories about globalization (Greeven et al., 2019; Yip et al., 2000). However, despite the discourse on the concepts and indicators of globalization, the lack of application of Yip's globalization driver framework warrants research on its usage.

In theory, as mentioned above, the drivers have been identified and underlined as the necessary conditions in an industry to create an ecosystem that can go international or become more global (Kudina et al., 2008). In other words, the drivers known as the "industry globalization drivers," were part of an organization's external environment rather than intrinsic or internal. The four drivers include market, cost globalization, competitive, and government drivers. An industry possessing all four drivers creates the potential for both MSMEs and large corporations, such as multinational corporations (MNCs), to achieve global strategy benefits. This concentration means the industry fosters an atmosphere that creates and supports an MSME's growth potential. The growth of MSMEs is a phenomenon widely studied in the literature. Therefore, this framework can add to a growing list of factors influencing the growth of MSMEs. Unlike KOF, MGI, and NGI indices that focus more on macroeconomics, Yip's framework leans more toward microeconomic causes, which is better suited to studying the growth of small businesses. However, these drivers are not without their issues.

Market Drivers

Under the market driver dimension, the emphasis is on dealing with consumer behavior at the macroeconomic level and the structure of distribution channels (Yip et al., 1997). The rise of globalization means a steady increase in the need for uniform products (Medina & Duffy, 1998). Because globalization idealizes a unified world in economic and social aspects, customers will be demanding similar products and services rather than products that are different and unique from one another or heterogeneous. These homogenous products offer the opportunity for organizations to conduct economies of scale, whereby costs can be reduced with every increase in output. This activity is achieved by standardizing products and services to cater to consumer demand, becoming homogenous or similar (Quelch & Deshpande, 2004). Standardization allows organizations to cut costs by using the same production and selling methods and saving costs from having outlets and branches with different cultures or management styles to adapt to local scenarios.

In reality, however, the diverse nature of cultures and attitudes of consumers from different areas of the world means that a degree of customization of products is still applicable and crucial (Zhang et al., 2015). Geographically, seen that one particular location has its tastes and preferences (Cortright, 2002). For instance, tastes and preferences in Asia often differ from those in Africa. One prime example of standardization while offering a few customizations is McDonald's, which can attribute its success to standardized methods and operations while also adapting to local tastes and preferences by giving the option to eat local foods.

In terms of distributing the products, a uniform global distribution channel can be established to satisfy just about any customer. Apart from standardizing products, other aspects that could be standardized include branding, advertising, and marketing of products. However, the main issue with this driver is that it does not allow organizations to create value via the differentiation of products. This is because it would be difficult to find opportunities to differentiate products from competitors, and lack of differentiation would discourage organizations, especially MSMEs, from innovating to promote healthy competition in an industry. In addition, however, uniformed products would encourage corporations to charge high prices as consumers flock to competitors offering similar yet cheaper products (Aapaoja & Haapasalo, 2014).

Cost Drivers

Cost globalization drivers focus on two aspects: economies of scale (Ambrose et al., 2019) and economies of scope (Akerman, 2018). Standardizing goods and services allows cost reduction from large amounts or scope of output. In addition, the scale at any location can be made larger by participating in multiple markets from standardized products.

Economies of scale and scope can be achieved by bulking. This is known as the purchasing economy, whereby bulk-buying of products will result in discounts (Büchi et al., 2018). In this case, a consolidation between ventures to gain purchasing economies would result in more purchasing power, leading to even larger purchases for further lowering costs. When combined, the purchasing power could rival governments, and the bigger the assets held by a merger, the higher the potential discounts received.

Transportation logistics is also an area that can be devoted to economies of scale. Specifically, the value of sales revenue to transportation costs must be favorable for value creation. Other promising attributes of products include just-in-time delivery, close location to consumers, non-perishability or long-lasting, and the shape of products to allow easy packaging for delivery; all contribute to transportation economies of scale.

However, this type of economy of scale is prone to disruptions (Berechman & Giuliano, 1985); should one part of the market fail, the entire system is halted, hindering the production of standardized products. On the other hand, should a fall scenario in market participation, it is still possible to gain economies of scale via the accumulation of learning and experiences Field (Cantor & Hewlett, 1988); steep learning curves often provide more benefits in the long run. Thus, close coordination must be made between market participants. Furthermore, investing in automation or high-tech machinery could reduce human error and place more emphasis on capital-intensive production, potentially increasing economies of scale.

Government Drivers

Arguably, the most important driver involves the government's intervention at the macroeconomic level. The trade policies Field (Baffes & Gorter, 2022) is an important factor influencing business transactions. Trade Protectionist policies that hinder trade include tariffs, import duties, quotas, embargoes, and export subsidies that can discourage trade from flourishing in an industry (Gezim & Bashkim, 2019). At the same time, these policies can help present local products and exports with the chance to compete against imports, resulting in less trade and reduced economic growth due to the decline of international trade brought by the lack of imported goods into the economy.

These policies consist of rules and regulations organizations must adhere to, and different governments impose different trade policies. However, a favorable trade policy promotes free trade to remove barriers. However, government drivers involve more than just trade policies. Government drivers also embody marketing regulations and technical standards implemented by government policymakers (Kudina et al., 2008; Yip et al., 2000).

While uniform global marketing is the preferred approach to save costs, each country has a different marketing environment or ecosystem (Sheth & Parvatiyar, 2001, p. 200). Marketing regulations in the United States are typically more liberal than those in Asia Field (Compton et al., 2011), especially on television advertising content. While globalization promotes the idea of one entity, the difference in technical standards still needs to improve the possibility of standardized products. The more restrictions being imposed on trade, the lesser the case for a standardized product.

Despite the importance of government drivers due to the power wielded by governments, organizational roles are becoming just as important. In the past, MNCs had to abide by whatever rules were set out; in today's context, however, a new trend has emerged that some MNCs possess the same buying power or even surpass the purchasing power parity of governments (Joseph, 1999). This allows them to participate at political, economic, and social levels, influencing trade policies.

Despite international calls for free trade, still exist barriers everywhere. Therefore, organizations should consider trading blocs and look at their particular rules and regulations (Jacks & Novy, 2020). The European Union (EU) and the Association of Southeast Asian Nations (ASEAN) are examples of trading blocs. The main drawback of trading blocs is that they often only favor organizations within their sphere of influence; external organizations are subjected to unfavorable conditions such as high taxes and bureaucracy (Aziz et al., 2018). Despite the Asian region being less liberal than others, growth potential is highest in Asia. However, this is more because the economies are still emerging and the markets are untapped rather than liberal reasons (You, 1999).

Competitive Drivers

In the above discussion, the chapter mentioned that MNCs could hold enough power to influence external factors. This is especially true when competitors compete at the international level. Sustained efforts from said competitors are enough to bring about change, such as consumer tastes and preferences.

Competitive Drivers are actions and activities done by organizations with far-reaching macroeconomic or international consequences, capable of influencing competitors and countries. Competitive drivers are different in that they are malleable and can dynamically change, while markets, costs, and government drivers are essentially fixed for an industry. According to this dimension, competitors can have a minor role in influencing the other three drivers via sustained efforts.

The first perspective on the competitive drivers is from a first-mover standpoint (Schlie & Yip, 2000). Competitors can create a system of competitive interdependence among countries through sharing activities or actions. Consider a scenario when a competitor was composing killer ads claiming they are the leading organization in a particular industry (Söderlund & Dahlén, 2010). These activities would no doubt provoke reactions from competitors through increased market participation and uniform marketing to ensure their positions are not weakened. Any changes in scale and costs would no doubt affect the competitive parts due to shared activities. In addition, consumers may view market position in a host or lead the country as an indicator of overall quality. The automobile industry is a prime example involving competitive drivers; because the sharing of information amongst competitors can lead to lower costs, resulting in increased competitive interdependence (Banerji & Sambharya, 1998; Burgers et al., 1993).

The second perspective of competitive drivers involves reacting by matching and pre-empting individual moves made by competitors (Singh & Yip, 2000). Because of the reaction to a particular competitor, there is increased market participation and reactions, including standardizing products.

However, care must be taken; do not go above and beyond by establishing a cartel by merging that is considered illegal (Bolotova, 2009). Cartels are notorious for being a collection of colluding businesses that have a hidden agenda of manipulating prices for their gain at consumer expense. This is done by establishing an agreement with other competitors for setting a pre-determined price that is prohibitively high.

External Factors Influencing Growth

This chapter extends the model to address the issues left out by adding the factors influencing the growth of MSMEs in terms of sales, number of employees, and assets that are often cited in the literature. As outlined by George Yip's framework, the existence of favourable conditions highlights how organizations can flourish at the macroeconomic level. Moreover, the external factors influencing growth also emphasize how MSMEs can thrive at the superficial microeconomic level. Thus, it would be natural to synthesize the two theories to form a cohesive or holistic approach to what constitutes the growth of MSMEs from an extrinsic standpoint. Figure 1 showcases the proposed synthesized framework of favorable conditions for growth.

Figure 1. A Synthesized Framework of Favourable Conditions for Growth of MSMEs

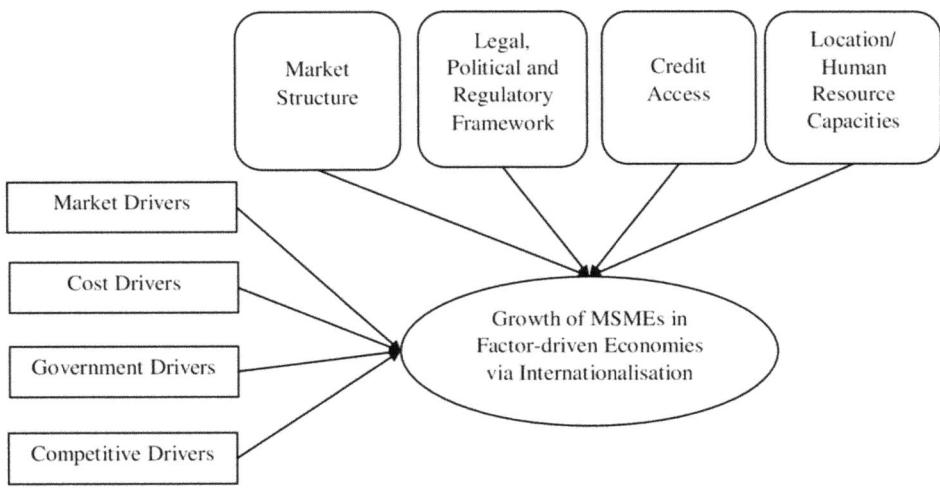

Market Structure

Market structure is one of the most significant external elements. The market structure Field determines the degree of competition and the number and percentage of businesses competing in an industry (G Djolov, 2014). Perfect competition, monopolistic competition, oligopoly, and monopoly are the four basic market structures that typically occur in an industry, and they are ranked from most competitive to least competitive (Al-Muharrami et al., 2006). Perfect competition has most businesses competing in an industry, whereas monopoly has just one firm.

Suffice it to say that MSMEs will likely survive in a competitive climate (Sun & Yannelis, 2007). Competition in a sector is sometimes regarded as beneficial to economic progress since it fosters product innovation and avoids pricing discrimination. These marketplaces have very low barriers to entry in many industries that are either monopolistically competitive or function under conditions that are close to perfect competition. This indicates that even a huge number of enterprises exist, each with a small market share. Each business with a greater firm size is more likely to be small than the total market size. This is exacerbated by the emergence of the Internet and new technology, dramatically decreasing startup costs and obstacles to any contestable business entry (Buss, 2002).

However, perfect competition is no guarantee that MSMEs will succeed. For example, in a perfectly competitive market owing to market saturation, even a reasonably strong demand would not provide reasonable profit, prompting the MSME to close down due to average costs exceeding average income. However, one study contradicted this by citing rivalry as one of the biggest impediments to the success of small businesses (Amankwah-Amoah et al., 2018).

Because a single entity holds price fixing, an industry monopolized by a single corporation has a high barrier to entry (Carare, 2011). It is commonly held that monopoly is ultimately the most harmful to the growth of MSMEs. The high obstacles to entry for MSMEs and the high customer switching costs are enough to dissuade even the most committed entrepreneurs (Campos et al., 2010). Monopolies, like mergers and cartels, frequently force smaller businesses out of business. When one corporation dominates a certain market, it can set product prices. Because most monopolizing corporations are so enormous, they can afford to decrease their prices so that no small firm can compete. As a result, smaller enterprises are forced to shut down or combine with monopolizing companies.

Some say monopolies are advantageous because highly lucrative corporations invest more in research and development (Hennipman, 1954). Because of its strong position, the monopoly can safely handle the risks involved with innovation. A very successful trust, on the other hand, may need more motivation to improve as long as consumers continue to exhibit a demand for their current product or service. On the other hand, businesses in a competitive market can compete by making improvements to present products and services and cutting pricing. Monopolies ensure high entry barriers and no free riding or modifications to their existing patents. A monopolized industry's workforce may also be much less than a competitive industry (Djolov, 2014). The procedures and expenses of launching a firm are time-consuming and bureaucratic, resulting in stronger and more fair competition and a negative impact on investments.

Legal, Regulatory, and Political Framework

The legal, regulatory, and political frameworks are the administrative agencies of a government that can conduct control and orders through policies. They are ultimately accountable for the development and nature of the business environment in the economy (Britton-Purdy et al., 2019). Examples include interest rates, rules and regulations, government infrastructure spending, the taxation system, and the circulation of money supply in an economy; thus, it may potentially alter the growth of businesses (Mokhova & Zinecker, 2014).

It is observed that the business climate or environment impacts the growth of MSMEs (Lumpkin & Dess, 1996). This comprises the legislative and regulatory framework of an economy that determines an ecosystem, which affects MSMEs competing in the ecosystem. Therefore, an unfavorable business climate has a detrimental impact on the growth of small businesses. While competition is regarded as healthy and beneficial since it stimulates organizational innovation and the avoidance of monopoly and dominating enterprises, it has been noted that competition may be a key impediment to the growth of the MSMEs (Brown, 2007).

Taxation is important in deciding entrepreneurship from a macroeconomic standpoint. An unfavorable tax structure, such as a high corporation tax, might stifle the expansion of the MSMEs (Ameyaw et al., 2016). Due to their small organizational size and lack of equality, MSMEs will be discouraged from hiring staff if income taxes are raised. There have been situations where the government favors the development of monopolies, particularly when regular delivery of a product or service with a relatively high up-front cost is required. Electric and water utilities are examples of such goods since it is frequently expensive to develop new power plants or dams; thus, to remain economically viable, monopolies must be permitted to regulate pricing to cover these expenses.

Furthermore, rules and regulations governing entry into a sector must be enforced and softened to allow for the easy entry of MSMEs eager to compete. Finally, simple policies must be introduced to reduce bureaucracy and paperwork on startups since convoluted laws and regulations can severely impede the growth of small businesses.

Finally, economic corruption impacts entrepreneurial activity (Krasniqi, 2007). Therefore, one of the key causes of the growth in unfair competition is corruption. He also underlined that the cost of complying with laws and rising tax rates raises the expenditures of small businesses while limiting their expansion. Thus, the biggest barriers to small business growth are unfair competition from the informal sector, onerous laws, and tax rates.

According to a study conducted by the International Finance Corporation in 2013 (Basnett, 2017), based on responses from more than 45,000 firms in developing countries, the top obstacles to their operations are a poor investment climate, particularly red tape, high tax rates, and competition from the informal sector, as well as inadequate infrastructure, particularly with an insufficient or unreliable power supply. Therefore, this informal economy may be considered a key impediment for MSMEs in developing nations (Bonnet & Venkatesh, 2016; Skenderi et al., 2017).

As a result, this emphasizes the need for government macroeconomic intervention in the form of fiscal, monetary, and supply-side policies that have a long-term impact on an economy's entrepreneurial activity. According to macroeconomic rules, the more the government expenditure on infrastructure such as roads and buildings, as well as the circulation of money supply in an economy, the better the nurturing of an entrepreneurial environment. At the same time, an economy with a high taxation rate and high-interest rates on borrowing will have a detrimental impact on the growth of MSMEs.

Credit Access

MSMEs require essential financing to begin operations to construct their firm from the bottom up. Access to capital would assist them in starting a business and growing, prospering, and expanding their sales and firm-size operations. Furthermore, organizations with greater access to money are better positioned to capitalize on growth and investment possibilities (Beck & Demirguc-Kunt, 2006).

It has been shown that more finance is needed to ensure micro-firms' expansion (Mbugua et al., 2013). Lack of access to external funding is a key barrier to MSMEs' growth, and it has been blamed for high rates of failure among such MSMEs (Abdul Rahman et al., 2016). However, entrepreneurs with the best ideas and entrepreneurial traits can mobilize their resources and money if they obtain the appropriate financing to launch their MSME. Regrettably, financial institutions and banks generally are hesitant to lend to budding entrepreneurs.

This is true because many lenders need help to analyze the relevant risks of an investment, and the presence of a high degree of ambiguity and lack of confidence provides them with additional reasons not to lend. As a result, a business strategy is essential for obtaining financial backing. It is argued that regulations favor large organizations, whereas small businesses experience challenges and difficulties due to a lack of access to funding (Nichter & Goldmark, 2009).

Youth entrepreneurs, in particular, need help to obtain funds to invest in their firms, such as initial seed money, and to finance company development. However, due to a lack of self-sustaining resources, a lack of a substantial credit history, and insufficient collateral or guarantees to acquire loans or lines of credit, young individuals are sometimes viewed as particularly risky investments, making access to financing difficult (Sharu & Guyo, 2013).

Financial institutions are more cautious when lending to MSMEs and MSMEs are often charged comparably high-interest rates, collateral, and loan guarantees (Shah et al., 2013). In addition, loan regulations and collateral requirements prevent enterprises from receiving bank loans (Krasniqi, 2007).

MSMEs in developing and established economies have had reduced access to external funding (N. Berger & F. Udell, 1998). As a result, MSMEs are more limited in their operations and expansion compared to large enterprises. They have conducted research in developing nations proving that MSMEs suffer larger funding challenges than big enterprises (Beck & Demirguc-Kunt, 2006). Funding, criminality, and political instability directly impact small enterprises' expansion pace, with financing being the most important limitation to small company growth (Ayyagari et al., 2006).

A limitation on company expansion in developing countries (Rocha et al., 2011): they discover that each nation has a unique set of constraints, which vary according to firm characteristics, particularly firm size. On the other hand, access to funding appears to be among the most significant impediments across all nations, whereas other obstacles tend to matter far less.

To satisfy their credit demands, MSMEs require access to financial institutions that can offer them loans at cheaper rates and on more acceptable conditions than a typical money lender. Traditional sources of credit in the formal sector, such as public sector banks, have assisted these enterprises (Bellavitis et al., 2017).

Location and Human Resource Capacities

Choosing a location for a business has always been critical. Location is crucial in attracting and maintaining the finest workers, many of whom pay special attention to where they work to maximize work-life balance. Wise site selections can considerably boost a company's long-term success. Above all, the location where an MSME may easily access human resources. Human resource capacities are one of the most important components in the success of small businesses.

Firms with talented and well-educated personnel will likely be more efficient (Hewitt & Wield, 1992). Human resource capacities are among the most important factors influencing the success of MSMEs (Y. Lee et al., 2011). Human resource capacities favourably influence small enterprises' growth, increasing employee skills and motivation, and, as a result, boosting small company productivity and long-term sustainability (Chandler & Mcevoy, 2000). A well-educated and talented workforce has more learning and inventive capacities (Charoenrat. Various studies, however, identify insufficient human resource skills as a fundamental barrier to developing MSMEs in developing nations (Charoenrat et al., 2013; Y. Lee et al., 2011). This is due to a lack of knowledge and infrastructure, which prevents trained and valued human resources from being easily accessible. A lack of entrepreneurial knowledge is very harmful to a company.

In today's highly competitive and dynamic corporate contexts, entrepreneurial studies, courses, and majors are among higher education's most essential study subjects. Entrepreneurship courses frequently attempt to stimulate entrepreneurial aspirations among students, equip them with the necessary entrepreneurial knowledge and abilities, and instill in them the required confidence to survive in the real world after graduation. These courses, in particular, are intended to improve students' capacity to build original and distinctive ventures and enhance their abilities as effective and inventive managers, which will eventually help them manage and expand their businesses.

Entrepreneurial education differs from standard business degrees, including practical programs and academic teachings taught in seminars and lectures. These useful programs aim to surpass the usual emphasis on learning ideas and case studies. Fundamentally, in today's complicated world, students must be able to 'practically' make things happen rather than merely intellectually. In addition, there is a rising shortage of employment, and students must increasingly have the essential practical skills to generate careers for themselves.

While there has been an ongoing discussion for decades over whether students can be educated to be entrepreneurs regardless of how sophisticated the education system is (Pache & Chowdhury, 2012; Varadarajan Sowmya et al., 2010), it cannot be denied that entrepreneurial education has become a mainstay in many academic institutions throughout the world. Scholars and academics have concluded that certain favorable traits, such as practical programs, business planning activities, interactive aspects, and integrated feedback mechanisms, distinguish good entrepreneurship courses (Warhuus et al., 2018). However, like any other course at an academic institution, the age-old assertion and dispute that there is a large divide between academics and practitioners persist (Dostaler & Tomberlin, 2013). The purpose of practical programs in entrepreneurship education is to equip students with real-life practical experiences to bridge the gap between theory and practice as much as feasible.

As a result, coaching is one approach to increase the quality of these practical programs in entrepreneurship education (Ben Salem & Lakhal, 2018). In certain circumstances, students need more personal coaching to engage in business activities. Thus, coaching provides a solution for improving practical programs and avoiding certain challenges caused by a lack of human connections between educators and students. Such human relationships are required, more so in functional scenarios than in classes, because it is here that pupils need the most direction and instruction. Coaching will raise the efficacy of useful programs and the possibility of students completing their specified assignments, resulting in more proficient and better-trained pupils. As a result, coaching applications must be included in such practical programs.

SIGNIFICANCE OF RESEARCH

Considering the above deliberation, the justification for this study is glaring; in a globalized economy, small businesses found it increasingly difficult to compete as it is now easier for large corporations to enter the international territory, despite the efforts of government intervention such as tariffs, quotas, and embargoes to support local products. With this research, small businesses can refer to the conditions suitable for their expansion and could devote their limited resources to a particular area. Another minor objective of this chapter is to clarify and explain the origins of globalization. This is because misconceptions regarding globalization are still widespread. Another research gap is the need for a framework to showcase drivers, yet these need to be specifically tailored for small businesses as they are too generalized.

Implications

This chapter will aid MSMEs in emerging economies since it will serve as a guideline for how to build their businesses to survive in today's unpredictable, competitive, and dynamic business climate. The policy conclusion is that governments may enact MSME-friendly laws by identifying the external elements that favor MSMEs' growth to build an entrepreneurial environment and boost economic growth. Policies concentrating on external issues, particularly, are critical to promoting MSMEs' development, which means more revenue for the government.

Limitations and Future Research Directions

However, this chapter is conceptual and thus does not focus on testing the hypotheses of any research objectives. Furthermore, this chapter did not include internal factors of MSMEs that may affect their performance at the international level. Future researchers can focus on empirical papers that verify the conceptual framework in Figure 1 while also considering the internal factors influencing the growth of

MSMEs from the literature. Emphasis must be placed on MSMEs located in least-developed economies, as these low-income nations face serious structural hurdles to long-term growth, have low levels of human resources, and are sensitive to economic and environmental shocks.

CONCLUSION

Whether it is favored or otherwise, organizations everywhere are inevitably part of the global markets. Economic and social consequences brought about by globalization would no doubt affect MSMEs and large corporations alike. Organizations everywhere must possess a degree of malleability to be dynamic and flexible enough to change the environment rapidly. Management and the workforce must have the skills, knowledge, willingness, and capabilities to change, though this falls under the category of change management. Fostering a dynamic ecosystem and approach to prepare, support, and help individuals, teams, and organizations make organizational changes would no doubt leave organizations prepared to face challenges should the four globalizations mentioned above drivers be non-existent in an industry.

The purpose of this chapter was to showcase a background and overview of globalization and how favorable globalization, including market drivers, cost globalization drivers, competitive drivers, and government drivers conditions, are beneficial for the growth of MSMEs and can also cause problems. This chapter also introduced an expanded George Yip's framework of globalization that showcases the necessary conditions that allow MSMEs to grow and flourish, which include external factors such as market structure, legal, regulatory, and political framework, credit access, location, and human resources capacities.

REFERENCES

Aapaoja, A., & Haapasalo, H. (2014). *The Challenges of Standardization of Products and Processes in Construction*. Research Gate.

Abdul Rahman, N., Yaacob, Z., & Mat Radzi, R. (2016). The Challenges Among Malaysian SMEs: A Theoretical Perspective. *World Journal of Social Sciences*, *6*, 124–132.

Akerman, A. (2018). A theory on the role of wholesalers in international trade based on economies of scope. *The Canadian Journal of Economics. Revue Canadienne d'Economique*, *51*(1), 156–185. doi:10.1111/caje.12319

Al-Muharrami, S., Matthews, K., & Khabari, Y. (2006). Market structure and competitive conditions in the Arab GCC banking system. *Journal of Banking & Finance*, *30*(12), 3487–3501. doi:10.1016/j.jbankfin.2006.01.006

Amankwah-Amoah, J., Antwi-Agyei, I., & Zhang, H. (2018). Integrating the Dark Side of Competition into Explanations of Business Failures: Evidence from a Developing Economy: Integrating the Dark Side of Competition into Explanations of Business Failures. *European Management Review*, *15*(1), 97–109. doi:10.1111/emre.12131

Ambrose, B. W., Fuerst, F., Mansley, N., & Wang, Z. (2019). Size effects and economies of scale in European real estate companies. *Global Finance Journal*, *42*, 100470. doi:10.1016/j.gfj.2019.04.004

Ameyaw, B., Korang, J., Twum, E., & Asante, I. (2016). Tax Policy, SMES Compliance, Perception and Growth Relationship in Ghana: An Empirical Analysis. British Journal of Economics. *Management & Trade*, *11*(2), 1–11. doi:10.9734/BJEMT/2016/22030

Anand, D. M. (2015). *Globalization and Indian School Education: Impact and Challenges*. CORE.

Asgary, N., & Walle, A. H. (2002). The cultural impact of globalization: Economic activity and social change. *Cross Cultural Management*, *9*(3), 58–75. doi:10.1108/13527600210797433

Ayyagari, M., Kunt, A. D., & Maksimovic, V. (2006). *How important are financing constraints?: The role of finance in the business environment*. World Bank Publications.

Aziz, N., Hossain, B., & Mowlah, I. (2018). Does the quality of political institutions affect intra-industry trade within trade blocs? The ASEAN perspective. *Applied Economics*, *50*(33), 3560–3574. doi:10.1080/00036846.2018.1430336

Baffes, J., & Gorter, H. (2022). *Experience With Decoupling Agricultur Al Support*. Social Sciences Research Network.

Banerji, K., & Sambharya, R. B. (1998). Effect of network organization on alliance formation: A study of the Japanese automobile ancillary industry. *Journal of International Management*, *4*(1), 41–57. doi:10.1016/S1075-4253(98)00003-9

Basnett, Y. (2017). What do empirical studies say about economic growth and job creation in developing countries? Beck, T., & Demirguc-Kunt, A. (2006). Small and medium-sized enterprises: Access to finance as a growth constraint. *Journal of Banking & Finance*, *30*(11), 2931–2943. doi:10.1016/j.jbankfin.2006.05.009

Bellavitis, C., Filatotchev, I., Kamuriwo, D. S., & Vanacker, T. (2017). Entrepreneurial finance: New frontiers of research and practice. *Venture Capital, 19*(1–2), 1–16. do i:10.1080/13691066.2016.1259733

Ben Salem, A., & Lakhal, L. (2018). Entrepreneurial coaching: How to be modeled and measured? *Journal of Management Development, 37*(1), 88–100. doi:10.1108/ JMD-12-2016-0292

Berechman, J., & Giuliano, G. (1985). Economies of scale in bus transit: A review of concepts and evidence. *Transportation, 12*(4), 313–332. doi:10.1007/BF00165470

Berger, N., A., F., & Udell, G. (1998). The economics of small business finance: The roles of private equity and debt markets in the financial growth cycle. *Journal of Banking & Finance, 22*(6), 613–673. doi:10.1016/S0378-4266(98)00038-7

Block, D. (2004). Globalization, Transnational Communication, and the Internet. *International Journal on Multicultural Societies, 6*.

Bolotova, Y. V. (2009). Cartel overcharges: An empirical analysis. *Journal of Economic Behavior & Organization, 70*(1), 321–341. doi:10.1016/j.jebo.2009.02.002

Bonnet, F., & Venkatesh, S. (2016). *Poverty and Informal Economies.* Oxford University Press. https://halshs.archives-ouvertes.fr/halshs-01297260

Britton-Purdy, J., Grewal, D. S., Kapczynski, A., & Rahman, K. S. (2019). Building a Law-and-Political-Economy Framework: Beyond the Twentieth-Century Synthesis. *The Yale Law Journal, 129*, 1784.

Brown, R. (2007). *Promoting entrepreneurship in arts education.* doi:10.4337/9781848440128.00017

Büchi, G., Cugno, M., & Castagnoli, R. (2018). Economies of Scale and Network Economies in Industry 4.0. Theoretical Analysis and Future Directions of Research. Symphonya. *Emerging Issues in Management, 2*(2), 6. doi:10.4468/2018.02.01

Burgers, W. P., Hill, C. W. L., & Kim, W. C. (1993). A theory of global strategic alliances: The case of the global auto industry. *Strategic Management Journal, 14*(6), 419–432. doi:10.1002mj.4250140603

Buss, D. D. (2002). Technology in the Internet age. *2002 IEEE International Solid-State Circuits Conference. Digest of Technical Papers (Cat. No.02CH37315),* (pp. 18–21). IEEE. 10.1109/ISSCC.2002.992920

Campos, N. F., Estrin, S., & Proto, E. (2010). *Corruption as a Barrier to Entry: Theory and Evidence* (*SSRN* Scholarly Paper ID 1711074). Social Science Research Network. https://papers.ssrn.com/abstract=1711074 doi:10.2139/ssrn.1693340

Cantor, R., & Hewlett, J. (1988). The economics of nuclear power: Further evidence on learning, economies of scale, and regulatory effects. *Resources and Energy*, *10*(4), 315–335. doi:10.1016/0165-0572(88)90009-6

Carare, P. M. (2011). *Monopoly: Advantages and Disadvantages* (*SSRN* Scholarly Paper ID 1787089). Social Science Research Network. doi:10.2139/ssrn.1787089

Ch, M. A., Faheem, M. A., Dost, M. K. B., & Abdullah, I. (2011). Globalization and its *Impacts on the World Economic Development, 2*(23), 8.

Chandler, G. N., & Mcevoy, G. M. (2000). Human Resource Management, TQM, and Firm Performance in Small and Medium-Size Enterprises. *Entrepreneurship Theory and Practice*, *25*(1), 43–58. doi:10.1177/104225870002500105

Charoenrat, T., Harvie, C., & Amornkitvikai, Y. (2013). Thai manufacturing small and medium-sized enterprise technical efficiency: Evidence from firm-level industrial census data. *Journal of Asian Economics*, *27*, 42–56. doi:10.1016/j.asieco.2013.04.011

Chew, S. C., Robertson, R., & Garrett, W. R. (1993). Review of Globalization: Social Theory and Global Culture.; Religion and Global Order., Roland Robertson [Review of Review of Globalization: Social Theory and Global Culture.; Religion and Global Order., Roland Robertson, by R. Robertson & W. R. Garrett]. *Contemporary Sociology*, *22*(6), 828–830. doi:10.2307/2075975

Coe, D. T. (2007). *Globalization and Labour Markets: Policy Issues Arising from the Emergence of China and India*. OECD. doi:10.1787/1815199X

Compton, R. A., Giedeman, D. C., & Hoover, G. A. (2011). Panel evidence on economic freedom and growth in the United States. *European Journal of Political Economy*, *27*(3), 423–435. doi:10.1016/j.ejpoleco.2011.01.001

Cortright, J. (2002). The Economic Importance of Being Different: Regional Variations in Tastes, Increasing Returns, and the Dynamics of Development. *Economic Development Quarterly*, *16*(1), 3–16. doi:10.1177/0891242402016001001

Crafts, N. (2004). Globalization and Economic Growth: A Historical Perspective. *World Economy*, *27*(1), 45–58. doi:10.1111/j.1467-9701.2004.00587.x

David, A., & Barwinska-Malajowicz, A. (2018). Where do We Go from Here? The EU Migration Flows after the Brexit Referendum. Possible Future Scenarios in the Polish Example. *Journal of Globalization Studies*, *9*(2), 3–17. doi:10.30884/jogs/2018.02.01

Djolov, G. G. (2014). The Economics of Competition. *The Race to Monopoly*. Taylor and Francis. https://www.taylorfrancis.com/books/mono/10.4324/9781315785561/economics-competition-george-djolov

Dostaler, I., & Tomberlin, J. (2013). The great divide between business schools research and business practice. *Canadian Journal of Higher Education*, *43*(1), 115–128. doi:10.47678/cjhe.v43i1.1895

Dreher, A., Gaston, N., & Martens, P. (2008). *Measuring Globalization*. Springer New York., doi:10.1007/978-0-387-74069-0

Farzanegan, M. R., Feizi, M., & Gholipour, H. F. (2021). Globalization and the Outbreak of COVID-19: An Empirical Analysis. *Journal of Risk and Financial Management*, *14*(3), 105. doi:10.3390/jrfm14030105

Gezim, J., & Bashkim, B. (2019). Trade Barriers and Exports between Western Balkan Countries. *Naše Gospodarstvo/Our Economy, 65*(4), 72–80.

Goldstein, N. (2010). *Globalization and Free Trade*. Infobase Publishing.

Greeven, M. J., Yip, G. S., & Wei, W. (2019). *Pioneers, Hidden Champions, Changemakers, and Underdogs: Lessons from China's Innovators*. MIT Press. doi:10.7551/mitpress/12007.001.0001

Gygli, S., Haelg, F., Potrafke, N., & Sturm, J.-E. (2019). The KOF Globalisation Index – revisited. *The Review of International Organizations*, *14*(3), 543–574. doi:10.100711558-019-09344-2

Hadidi, H. E., & Kirby, D. A. (2015). Universities and Innovation in a Factor-Driven Economy: The Egyptian Case. *Industry and Higher Education*, *29*(2), 151–160. doi:10.5367/ihe.2015.0248

Hax, A. C. (1989). *Building The Firm Of The Future*. ProQuest. https://www.proquest.com/openview/63c2ebf36830d8c46ee32fc7482c4ecc/1?pq-origsite=gscholar&cbl=26142

Heimberger, P. (2020). Does economic globalization affect income inequality? A meta-analysis. *World Economy*, *43*(11), 2960–2982. doi:10.1111/twec.13007

Hennipman, P. (1954). Monopoly: Impediment or Stimulus to Economic Progress? In E. H. Chamberlin (Ed.), *Monopoly and Competition and their Regulation: Papers and Proceedings of a Conference held by the International Economic Association* (pp. 421–456). Palgrave Macmillan UK. 10.1007/978-1-349-08434-0_22

Hewitt, T., & Wield, D. (1992). In T. Hewitt, D. Wield, & H. Johnson (Eds.), *Technology and Industrialization* (pp. 201–221). Oxford University Press. https://www.amazon.co.uk/Industrialization-Development-Tom-Hewitt/dp/0198773323

Hoang, H., & Antoncic, B. (2003). Network-based research in entrepreneurship: A critical review. *Journal of Business Venturing*, *18*(2), 165–187. doi:10.1016/S0883-9026(02)00081-2

Jacks, D. S., & Novy, D. (2020). Trade Blocs and Trade Wars during the Interwar Period. *Asian Economic Policy Review*, *15*(1), 119–136. doi:10.1111/aepr.12276

James, H. (2009). The End of Globalization: Lessons from the Great Depression. Harvard University Press.

Jamil, M. I. M., & Almunawar, M. N. (2021). Importance of Digital Literacy and Hindrance Brought About by Digital Divide [Chapter]. Encyclopedia of Information Science and Technology, Fifth Edition. IGI Global. doi:10.4018/978-1-7998-3479-3.ch116

Joseph, S. (1999). Taming the Leviathans: Multinational Enterprises and Human Rights. *Netherlands International Law Review*, *46*(2), 171–203. doi:10.1017/S0165070X00002394

Krasniqi, B. (2007). Barriers to entrepreneurship and SME growth in transition: The case of Kosova. [JDE]. *Journal of Developmental Entrepreneurship*, *12*(1), 71–94. doi:10.1142/S1084946707000563

Kudina, A., Yip, G., & Barkema, H. (2008). Born Global. *Business Strategy Review*, *19*(4), 38–44. doi:10.1111/j.1467-8616.2008.00562.x

Lee, E., & Vivarelli, M. (2006). The social impact of globalization in the developing countries. *International Labour Review*, *145*(3), 167–184. doi:10.1111/j.1564-913X.2006.tb00016.x

Lee, Y., Lin, B.-W., Wong, Y.-Y., & Calantone, R. J. (2011). Understanding and Managing International Product Launch: A Comparison between Developed and Emerging Markets: Understanding And Managing International Product Launch. *Journal of Product Innovation Management*, *28*(s1), 104–120. doi:10.1111/j.1540-5885.2011.00864.x

Loretta, L., & Bang, S. (2015). Hyun, & Morales [Global Gentrifications: Uneven Development and Displacement. Policy Press.]. *E (Norwalk, Conn.)*, L.

Lumpkin, G. T., & Dess, G. G. (1996). Clarifying the Entrepreneurial Orientation Construct and Linking It To Performance. *Academy of Management Review*, *21*(1), 135–172. doi:10.2307/258632

Máliková, I. (2020). Impact of globalization on circular economy and sustainable development. *SHS Web of Conferences, 74*, 06018. 10.1051hsconf/20207406018

Mayer, J. (2002). Globalization, technology transfer, and skill accumulation in low-income countries. In *Globalization, Marginalization and Development*. Routledge. doi:10.4324/9780203427637.ch5

Mbugua, J. K., Mbugua, S. N., Wangoi, M., & Kariuki, J. O. O. & J. N. (2013). *Factors Affecting the Growth of Micro and Small Enterprises: A Case of Tailoring and Dressmaking Enterprises in Eldoret*. http://localhost:8080/xmlui/handle/123456789/9928

Medina, J. F., & Duffy, M. F. (1998). Standardization vs globalization: A new perspective of brand strategies. *Journal of Product and Brand Management*, *7*(3), 223–243. doi:10.1108/10610429810222859

Mokhova, N., & Zinecker, M. (2014). Macroeconomic Factors and Corporate Capital Structure. *Procedia: Social and Behavioral Sciences*, *110*, 530–540. doi:10.1016/j.sbspro.2013.12.897

Molle, F. (2007). Scales and Power in River Basin Management: The Chao Phraya River in Thailand. *The Geographical Journal*, *173*(4), 358–373. doi:10.1111/j.1475-4959.2007.00255.x

Narula, R., & Dunning, J. H. (2000). Industrial Development, Globalization, and Multinational Enterprises: New Realities for Developing Countries. *Oxford Development Studies*, *28*(2), 141–167. doi:10.1080/713688313

Nichter, S., & Goldmark, L. (2009). Small Firm Growth in Developing Countries. *World Development*, *37*(9), 1453–1464. doi:10.1016/j.worlddev.2009.01.013

Norman, P. (2011). *The Risk Controllers: Central Counterparty Clearing in Globalized Financial Markets*. John Wiley & Sons.

O'Bannon, C., Carr, J., Seekell, D. A., & D'Odorico, P. (2014). Globalization of agricultural pollution due to international trade. *Hydrology and Earth System Sciences*, *18*(2), 503–510. doi:10.5194/hess-18-503-2014

Pache, A.-C., & Chowdhury, I. (2012). Social Entrepreneurs as Institutionally Embedded Entrepreneurs: Toward a New Model of Social Entrepreneurship Education. *Academy of Management Learning & Education, 11*(3), 494–510. doi:10.5465/amle.2011.0019

Quelch, J. A., & Deshpande, R. (2004). *The Global Market: Developing a Strategy to Manage Across Borders.* John Wiley & Sons.

Ritzer, G. (2008). *The Blackwell Companion to Globalization.* John Wiley & Sons.

Rocha, R. de R., Farazi, S., Khouri, R., & Pearce, D. (2011). The Status of Bank Lending to SMEs in the Middle East and North Africa Region: The Results of a Joint Survey of the Union of Arab Bank and the World Bank (*SSRN* Scholarly Paper ID 1794912). Social Science Research Network. https://papers.ssrn.com/abstract=1794912 doi:10.1596/1813-9450-5607

Schlie, E., & Yip, G. (2000). Regional follows global: Strategy mixes in the world automotive industry. *European Management Journal, 18*(4), 343–354. doi:10.1016/S0263-2373(00)00019-0

Shah, S. F. H., Nazir, T., Zaman, K., & Shabir, M. (2013). Factors Affecting the Growth of Enterprises: A Survey of the Literature from the Perspective of Small- and Medium-Sized Enterprises. *Journal of Enterprise Transformation, 3*(2), 53–75. doi:10.1080/19488289.2011.650282

Sharu, H., & Guyo, D. W. (2013). *Factors Influencing Growth of Youth Owned Small and Medium Enterprises in Nairobi County, Kenya., 4*(4), 8.

Sheth, J. N., & Parvatiyar, A. (2001). The antecedents and consequences of integrated global marketing. *International Marketing Review, 18*(1), 16–29. doi:10.1108/02651330110381952

Sifakis, N. C., & Sougari, A.-M. (2003). Facing the Globalisation Challenge in the Realm of English Language Teaching. *Language and Education, 17*(1), 59–71. doi:10.1080/09500780308666838

Singh, K., & Yip, G. S. (2000). Strategic Lessons from the Asian Crisis. *Long Range Planning, 33*(5), 706–729. doi:10.1016/S0024-6301(00)00078-9

Skenderi, N., Islami, X., & Mulolli, E. (2017). The Impact of Informal Economy in the Development of SMEs-Evidence from Kosovo (2008-2012). *International Journal of Management, Accounting, and Economics, 4*, 554–564.

Söderlund, M., & Dahlén, M. (2010). The "killer" ad: An assessment of advertising violence. *European Journal of Marketing*, *44*(11/12), 1811–1838. doi:10.1108/03090561011079891

Stahel, R. W., & MacArthur, E. (2019). *The Circular Economy: A User's Guide.* Taylor and Francis. https://www.taylorfrancis.com/books/mono/10.4324/9780429259203/circular-economy-walter-stahel-ellen-macarthur

Sun, Y., & Yannelis, N. C. (2007). Perfect competition in asymmetric information economies: Compatibility of efficiency and incentives. *Journal of Economic Theory*, *134*(1), 175–194. doi:10.1016/j.jet.2006.03.001

Tonts, M., & Siddique, M. B. (2011). *Globalization, Agriculture and Development: Perspectives from the Asia-Pacific.* Globalization, Agriculture and Development. https://www.elgaronline.com/view/edcoll/9781847208187/9781847208187.00005.xml

Tridico, P., & Paternesi Meloni, W. (2018). Economic growth, welfare models, and inequality in the context of globalization. *Economic and Labour Relations Review*, *29*(1), 118–139. doi:10.1177/1035304618758941

Urata, S. (2002). Globalization and the Growth in Free Trade Agreements. *Asia-Pacific Review*, *9*(1), 20–32. doi:10.1080/13439000220141569

Varadarajan Sowmya, D., Majumdar, S., & Gallant, M. (2010). Relevance of education for potential entrepreneurs: An international investigation. *Journal of Small Business and Enterprise Development*, *17*(4), 626–640. doi:10.1108/14626001011088769

Velasquez, M. (2000). Globalization and the Failure of Ethics. *Business Ethics Quarterly*, *10*(1), 343–352. doi:10.2307/3857719

Vujakovic, P. (2009). *How to Measure Globalization? A New Globalisation Index (NGI) (Working Paper No. 343).* WIFO Working Papers. https://www.econstor.eu/handle/10419/128904

Warhuus, J. P., Blenker, P., & Elmholdt, S. T. (2018). Feedback and assessment in higher-education, practice-based entrepreneurship courses: How can we build legitimacy? *Industry and Higher Education*, *32*(1), 23–32. doi:10.1177/0950422217750795

Watson, J. (2003). The potential impact of accessing advice on SME failure rates: The potential impact of accessing advice on SME failure rates. *Proceedings of the Small Enterprise Association of Australia and New Zealand 16th Annual Conference.* SEA.

White, R. D. (2010). *Global Environmental Harm: Criminological Perspectives.* Taylor & Francis.

Yameogo, C. E. W., Omojolaibi, J. A., & Dauda, R. O. S. (2021). Economic globalization, institutions and environmental quality in Sub-Saharan Africa. *Research in Globalization*, *3*, 100035. doi:10.1016/j.resglo.2020.100035

Yip, G. S., Biscarri, J. G., & Monti, J. A. (2000). The Role of the Internationalization Process in the Performance of Newly Internationalizing Firms. *Journal of International Marketing*, *8*(3), 10–35. doi:10.1509/jimk.8.3.10.19635

Yip, G. S., Johansson, J. K., & Roos, J. (1997). Effects of Nationality on Global Strategy. *MIR. Management International Review*, *37*(4), 365–385.

Yip George, S. (2004). Using Strategy to Change Your Business Model. *Business Strategy Review*, *15*(2), 17–24. doi:10.1111/j.0955-6419.2004.00308.x

You, J.-I. (1999). Income Distribution and Growth in East Asia. In *East Asian Development: New Perspectives*. Routledge.

Zhang, M., Lettice, F., & Zhao, X. (2015). The impact of social capital on mass customization and product innovation capabilities. *International Journal of Production Research*, *53*(17), 5251–5264. doi:10.1080/00207543.2015.1015753

KEY TERMS AND DEFINITIONS

Globalization: "Globalization" refers to the increasing interconnectedness of the world's economies, cultures, and inhabitants due to cross-border commerce in products and services, technology, and movements of investment, people, and information.

Internationalization: Internationalization refers to developing a product so that it may be easily consumed in many nations. Companies that want to expand their worldwide presence outside their home market employ this technique because they recognize that consumers in other countries may have different interests or habits.

Micro, Small, and Medium-sized Enterprises (MSMEs): On a small scale, small businesses or industries are those that create goods and services on a small scale. These industries are critical to a country's economic prosperity.

Section 2
Industry 5.0: CyberSecurity Essentials

Chapter 3

CyberSecurity Essentials for Industry 5.0

Mahmoud Numan Bakkar

https://orcid.org/0000-0003-4637-4035

Institute of Applied Technology, Abu Dhabi Vocational Education and Training Institute, UAE

ABSTRACT

Currently, hacking threats have increased exponentially because of the massive integration of technology into our daily life practices. Hackers are usually known for their advanced programming skills. They utilize these skills in challenging old systems and work on breaking them to test their capabilities and achieve their desire or motivation. The terminology of cybercrime evolved with the current industry's 4.0 and 5.0 revolutions and the changing of cybersecurity domains. This book chapter will discuss the different types of attacks in the industry 5.0 Era. Show examples of industrial cybersecurity attacks, Industry 5.0 cybersecurity vulnerabilities, and issues.

INTRODUCTION

Hacking threats have increased exponentially because of the massive integration of technology into our daily life practices. Hackers are usually known for their advanced programming skills. They utilize these skills in challenging old systems and work on breaking them to test their capabilities and achieve their desire or motivation. The terminology of cybercrime evolved with the current industry 4.0 and 5.0.0 revolutions and the changing of cybersecurity domains. The cybersecurity domain developed through the Internet enables companies to collect users' data such as names, backgrounds, friends, jobs, interests, travels, and locations. Internet

DOI: 10.4018/978-1-7998-8805-5.ch003

evaluations created many global domains for data, such as Google, for the fact that many people in the world use and have Google accounts, and it is available on 80% of mobile devices around the globe (Academy, 2022)

Facebook is also considered an enormous cybersecurity domain, fed with many personal data entered by users. In addition, the LinkedIn network of professional employees and employers is all connected through one network. Those domains, in addition to the industry's four technologies, such as the Internet of Things (IoT), Cyber-Physical Systems, Smart manufacturing, smart factories, cloud computing, and Artificial Intelligence, all created a demand for cybersecurity measures to protect the environment from being attacked and compromised for specific harmful actions. Industry 5.0 Cybersecurity Essentials emphasizes business continuity and risk management; more white and gray hat hackers are needed in the upcoming Industry 5.0. Having white hat hackers could reduce the vulnerability of the industrial systems; however, it may add more cost to the production line to create a fort line against the black hat hackers and the organized hackers, Script. Moreover, kiddies, black hackers, and organized hackers could have cyber motivation for their crimes, such as financial gain (Academy, 2022).

The first CyberSecurity Essential threat for Industry 5.0 is cyber criminals, who are either insiders such as employees, contractors, or outsiders such as hackers (black, white, grey) and organized attackers (Hacktivist, Terrorist, State-Sponsored). They are working towards financial gain; they can work on cracking passwords and sending malware and viruses to steal financial information, such as credit cards or any information that can be sold for high earnings. Cybercrime countermeasures varied based on data sensitivity and value. However, the current practices start with a system that alerts the victims early before the attack occurs using the honey bots and using shared knowledge and expertise of attacks, such as maintaining a vulnerability database that keeps the records of the national common vulnerability and exposures (CVE) (Academy, 2022)—also sharing intelligence, in addition, having laws information security management standards (ISM) such as ISO /IEC 27000 (Academy, 2022).

Databases are the primary concern of companies in the industry 5.0 Era. Medical, finance, education, and private and personal records can be the primary targets for cybercriminals. The second threat is the internet manufacturing infrastructure (IMI); many organizations use cloud computing technologies, so attacks such as DNS spoofing redirect the domain names to the attack's computer. Also, packet sniffing can be used to steal sensitive transmitted data such as credit cards, passwords, and attacks such as man in the middle can be used for packet forgery. In addition to attacking systems that control the manufacturing industry, for example, supervisory control and data acquisition (SCADA), an attack on industry needs such as telecommunication, logistical systems, transportation, electrical systems, and power providers. In the following chapter sections, we will: discuss the different types of attacks in the Industry 5.0 era; show examples of industrial cybersecurity attacks; and discuss Industry 5.0 cybersecurity vulnerabilities and issues; types of cybersecurity skills required for the Industry 5.0 era; protecting secrets and ensuring integrity; industrial approach for cybersecurity in Industry 4.0 and 5.0; information security standards and Frameworks; IT and OT concepts and cybersecurity in the Industry 5.0 era; and cybersecurity attack threats and their countermeasures for industry 5.0.

TYPE OF ATTACKS IN INDUSTRY 5.0 ERA

Social Engineering Attack: Cybercriminals can work on manipulating user data using their psychological behaviors and reactions, using different forms of social engineering attacks, such as baiting, based on promising victim-specific awards either online or physically using USP drives (Siddiqi et al., 2022). Shoulder surfing is another form where the criminal watches the victims while filling out forms or entering the PIN (Siddiqi et al., 2022). Pretexting requests client-sensitive data by impersonating an organization employee, such as a help desk (Academy, 2022). Phishing uses multiple emails and phone calls to request that the client provide sensitive data (Academy, 2022). Ransomware asks that victims pay ransom to return victim data that is held decrypted (Academy, 2022). Tailgating is by following the employee to go through the gate before closing (Academy, 2022). Finally, dumpster diving can be used to retrieve information from discarded information in trash bins or usernames, passwords left on yellow stickers, or electronically on desktops or USP drives (Siddiqi et al., 2022).

Scareware is a crucial attack in the industry 5.0 era, based on users' fear of installing fake antiviruses.

The Advanced Persistent Threat (APT) Attack occurs over a long period of time to seek' continuity until it achieves the target, which is either politically or business oriented motivated.

Federated identity attacks include users logging in to other websites using their Google credentials or their Facebook login. If the credentials are stolen, it will have a cascading effect; it is good to link the login with specific devices using the MAC address (Academy, 2022).

Denials of service attacks and distributed denial of service attacks, for example, can create numerous fake requests to the manufacturer's website, which blocks legitimate users from accessing the service online (Lacava et al., 2021).

Spoofing is used to steal data by impersonating other devices or users. It is used to force the robots, for example, to work differently or lose control of the drones, for example, by changing the GBS coordinates or the Internet Protocol (IP) or the Domain Name System (DNS) values (Lacava et al., 2021).

A man-in-the-middle attack, since many robotic systems nowadays use open-source code, and weak authorization and authentication, the man-in-the-middle attack, could occur if an attacker inserts malicious code or commands into the robotic system or communication channel used if it is not securely encrypted (Lacava et al., 2021).

Attacks were tampered with by changing configurations such as permissions or the parameters of the robots (Lacava et al., 2021).

Furthermore, there are five types of artificial intelligence agent models listed below:

1- simple reflex agent;
2- model-based reflex agent;
3- goal-based agents;
4- utility-based agents; and
5- learning agents.

As shown in one of the examples of the agents' models in Figure 1, all agent's model components are grouped into three types of mediums where cybersecurity attacks could occur:

Figure 1. Simple reflex agent
(Adapted from Stuart, 2010)

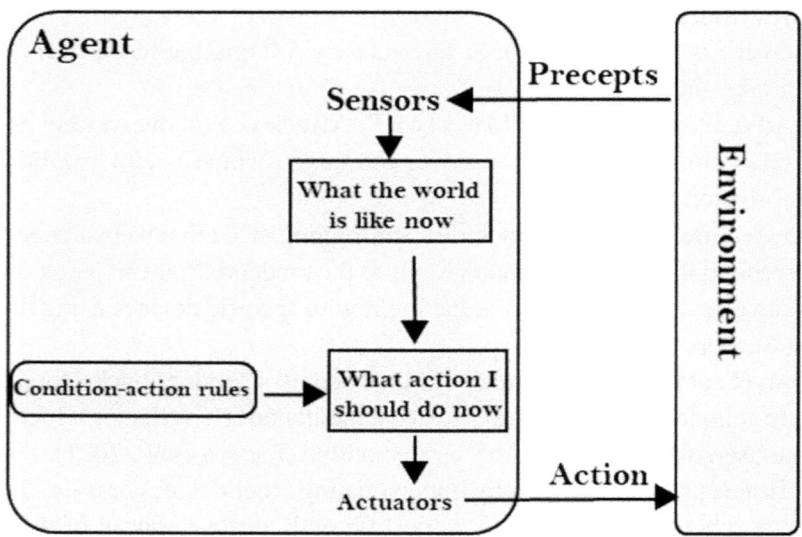

The three types of mediums are:

Hardware mediums such as sensors, actuators, display systems, and cameras: cybersecurity attacks such as phishing, and Trojans, which are used for back-door attacks that enable the attackers for unauthorized access to the AI agent;

Software mediums such as Python programs and firmware that control the agents: cybersecurity attacks such as worms, for example, Stuxnet attack, Trojans and random access Trojan (RAT) attacks such as Zeus (2007), Ransomware, which locks and encrypts the data to prevent unauthorized users from their access without paying by Bitcoin such as CryptoLocker (2007) (Yaacoub et al., 2021). Rootkit attacks target root access, such as the administrator's account level, to install a back door for future attacks, such as keeping spyware to attack the CIA attributes (Yaacoub et al., 2021). Botnet attack, which uses malware to control connected devices such as PCs, Servers, IoT devices, and mobiles, without being known by users, for example, Methbot (2016) (Yaacoub et al., 2021). Spyware attacks collect information about the Ai agent's activity and operation and send it to a third party. Buffering overflow attacks are used to attack AI agent memory. Password cracking, such as brute force attacks, is based on guessing, and dictionary attacks are based on a defined set of words (Yaacoub et al., 2021); and

Communication mediums such as Wi-Fi, Ethernet networks, Bluetooth devices, and cloud services: cybersecurity attacks such as jamming attacks are used to jam the communications between AI agents. De-authentication attacks are used to disable the current AI agent authentication and prevent it from re-connection. Eavesdropping attacks assisted in extracting passively sensitive information by monitoring the transmitted traffic between AI Agents' communications channels and the operators. Attacks on AI data integrity by injecting false data into the communication channel payload (Yaacoub et al., 2021). Distributed denial of service attacks on DDOS could also affect communication, preventing legitimate users from accessing AI agent services. Using a communication medium, the attacker could reveal the AI agent's identity and the operator's privacy in addition to the geographical location of the AI agent (Yaacoub et al., 2021).

EXAMPLES OF INDUSTRIAL CYBERSECURITY ATTACKS

Merck

The pharmaceutical company Merck was subjected to a Ransomware attack in 2017. The malware got into over 30,000 computers and 7,500 servers. Years of research were lost, and normal operations were seriously affected. It is unknown what the financial impact on Merck was, but the company ultimately sued its insurance providers, claiming $1.3 billion1 in losses (Boo, Patrik, n.d.).

Ukrainian Power

In 2015, a Ukrainian power producer was attacked by a team of hackers, likely a Nation-State. The attack led to over 225,000 customers2 losing power for 6 hours. The attackers used multiple strategies and methods to access the system and shut down power production. The critical factor is that the attackers had ample time to understand the system, embed themselves, and then wait for the right moment to strike (Boo, Patrik, n.d.).

Natanz Enrichment Facility

This case, commonly known as Stuxnet3, is one of the most famous cyberattacks ever in the industrial sector due to its sophistication and impact. It was one of the first malware files designed to attack an industrial system. The virus got into the facility and changed the running code in the SCADA system that controlled the centrifuges used to enrich uranium, with the intent of making them operate in a way that would lead to failures. As this occurred at a classified nuclear facility in Iran, the exact impact of the attack is unknown. One can only speculate about the extent of damage that could be inflicted in handling radio-active materials (Boo, Patrik, n.d.).

Flamer Virus

A virus infected the computers using Microsoft Windows operating systems in the Middle East region; it was used for recording Skype calls, keystrokes on the keyboards, and network traffics (Ervural & Ervural, 2018).

Maroochy Shire Council Sewage Spill in Queensland, Australia

It worked by failing the SCADA controller and all operations by controlling the wireless radio to dump the sewage water into the Maroochydore River in Queensland (Ervural & Ervural, 2018).

Steel Mill Attack in Germany

Using social engineering and spear-phishing, the attackers can put down all production machines by entering the factory network and the production network to manipulate the industrial controllers in the network (Ervural & Ervural, 2018).

Dam Attack in New York

Using a cellular modem, the attackers controlled the dam controller (Ervural & Ervural, 2018).

The three most significant attacks: The three most important attacks in the manufacturing industry are explained in the following text. The first one occurred in an Austrian aerospace company named FACC AG, which was working on manufacturing components for aerospace; the attack targeted the accounting department and their C-suite, the attackers requested the department to transfer 55.8 million for an acquisition deal, by impersonating the company CEO (Angel, 2022). Another attack, named Norsk Hydro, attacked Ransomware named LockerGoga and forced the company to close its operation in multiple countries such as Qatar, Norway, and Brazil. Moreover, it was estimated to cost the company around 75 million dollars (Angel, 2022). Finally, the automotive company Renault-Nissan was attacked using Ransomware called WannaCry, and a few employees were working (Angel, 2022). Ransomware stopped the company's production in five plants distributed in different countries (Angel, 2022).

INDUSTRY 5.0 CYBERSECURITY VULNERABILITIES AND ISSUES

In the Industry 5.0 era, many vulnerable devices and machines have been used exponentially, such as mobile devices and portable devices like: laptops and iPads; also, storage devices such as USP, SSD, and expanded storage devices use cloud computing. In addition, IoT devices use different types of sensors; big data infrastructure and robots.

Industry 5.0 will lead to massive, personalized designs of the products rather than only enormous production. This evolution in the industry will lead to more vulnerabilities and issues that the manufacturing stakeholders should address. First, different vulnerabilities could affect robotic systems, such as network ones that could be attacked with man-in-the-middle attacks, sniffing, and eavesdropping (Yaacoub et al., 2021). The second-mentioned vulnerabilities come from platforms that could attack because of the lack of patch updates (Yaacoub et al., 2021). Third, applications could also be affected by Cybersecurity attacks if they are not tested or evaluated appropriately (Yaacoub et al., 2021). We are not adopting good practices, planning, guidelines, or appropriate procedures, such as using incorrect security measures, tools, or lower coding skills, that produce less secure codes that can be easily attacked (Yaacoub et al., 2021). Finally, (Yaacoub et al., 2021) explained that heterogeneity and homogeneity vulnerability could attack the robotic systems because of the integration between robotic systems or the homogenous robotic features. Integrating robots with humans in Industry 5.0 will have issues related to the level of acceptance of robots in the organization's workplace, the changes in the organization's structure and its workflows, the organization's work ethics, the discrimination against people and robots, the level of trust and protecting the privacy within the human-robot collaboration environment, the workplace re-design to integrate the robots collaboration with humans, and the level of training and education needed in industry 5.0 environments (Demir et al., 2019). Industry 5.0 will require integrating technologies such as additive manufacturing, collaborative robots, the Internet of Everything, multi-agent systems and technologies, complex adaptive systems, smart manufacturing, digital ecosystems, and emergent artificial intelligence (Haleem & Javaid, 2019). The need for human collaboration with the above technologies could raise an issue of vulnerability if the people working moved to be a source of threats towards the industry 5.0 systems. So exceptional standards and frameworks must be developed in the future to handle this kind of threat.

TYPE OF CYBERSECURITY SKILLS REQUIRED FOR INDUSTRY 5.0 ERA

The current cybersecurity workforce operates based on the United States of America's national cybersecurity workforce framework; those skills are presented in Figure 2 (Brecht, 2020). The increased phenomena of the inter-connected physical-digital devices created more burdens on cybersecurity specialists to develop their security frameworks further and increase their capability to prevent any failure in the manufacturing environment. Starting with the CIA model, there are significant concerns about the user's confidentiality because they seek to protect their data privacy, and only authorized people can access sensitive data. Only access controls, data encryption, and authentication are insufficient in the industry 4.0 and 5.0 utilization processes. Hence demands meta-cognitive technology based on artificial intelligence and data science for deep analysis and prediction of any future occurrence of cyber attacks.

Moreover, countermeasures can be placed based on people's roles interacting with the system, users, or administrators' roles. For example, In a banking information systems, the people could be tellers, account managers, and customer service. Specific countermeasures should be placed among those people to avoid any security breaches. Further to people, policies and procedures should be implemented to manage the people and their interaction with the technologies used. Cybersecurity technology assists a lot nowadays in protecting the systems, such as: intrusion detection systems (IDS); intrusion prevention systems (IPS); firewalls; network scanners; antiviruses; and content filtering systems. Finally, despite the rapid development of technologies and the people's lack of skills related to up-to-date security attacks or approaches, the need for continuous awareness programs has been raised. This activity will reduce many security breaches.

The framework consists of seven layers of Cybersecurity skills required to cover different levels of Cybersecurity professions, starting from the lowest layer, investigation (IN). It needs the cybersecurity specialist to investigate any events or crimes in the IT environment of the industry 5.0 Technologies (Brecht, 2020). Specialists can have a profession such as incident response, vulnerability assessment, and management (Brecht, 2020). The second layer focuses on collecting and operating skills, demanding the specialist collect cybersecurity information for intelligence needs (Brecht, 2020). Finally, a specialist could have a profession that supports cyber defence infrastructure and covers investigative skills (Brecht, 2020).

After that, the specialist should be able to analyze any cybersecurity information for other helpful intelligence. Then, they can work to protect and defend by having cyber defence analysis skills and collecting and operating skills (Brecht, 2020). The next level of professionalism, which is needed to protect and defend, requires a high level of analysis skills; it requires the specialist to identify, analyze and mitigate any cyber threats (Brecht, 2020). Finally, the specialist could have a profession that requires all types and sources of analysis, such as language, exploitation, targets, and threat analysis (Brecht, 2020).

The fifth level of professions is required to operate and maintain by providing the needed support, administration, and maintenance to ensure the cybersecurity work in the organization runs efficiently. For example, a specialist could have a profession that requires operating and maintaining customer service, technical support, data administration, knowledge management, network services, system administration, and system analysis (Brecht, 2020). Finally, the sixth level of the

Figure 2. The National Initiative for Cybersecurity Education (NICE)
Adapted from (Brecht, 2020)

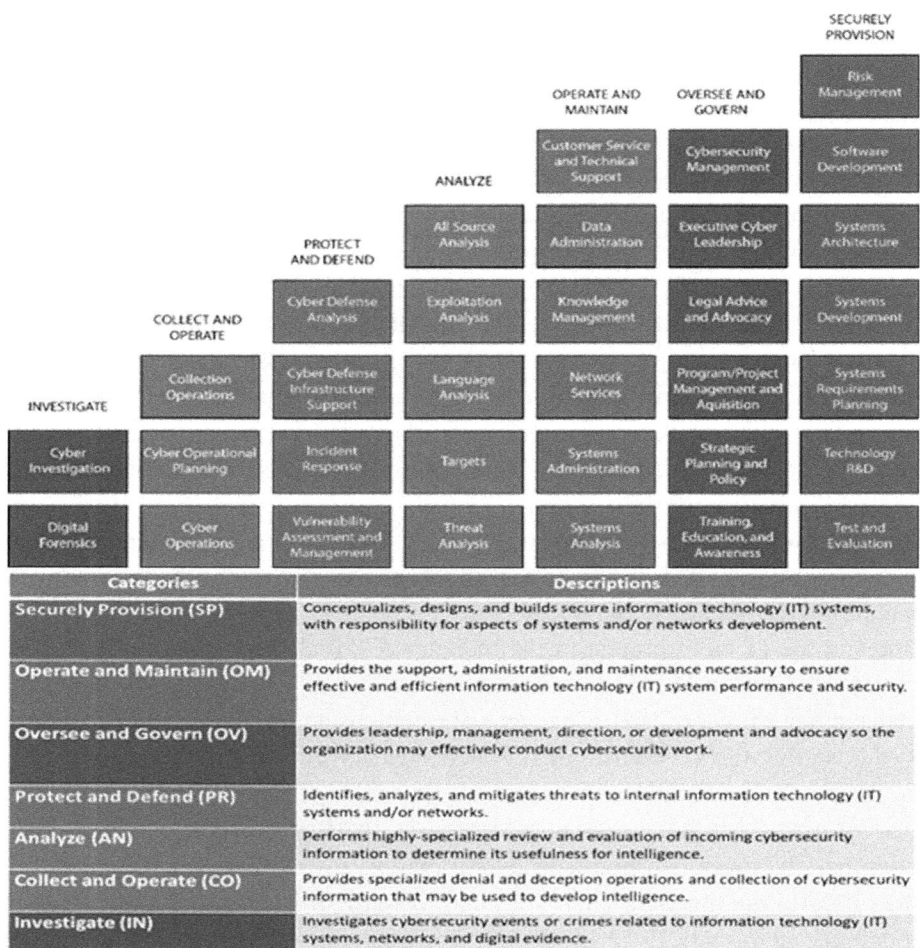

profession is required to oversee and govern by providing the needed leadership and management to effectively run the Cybersecurity work in the organization (Brecht, 2020). For example: a specialist could have a profession that oversees and governs Cybersecurity management; executive cyber leadership; legal advice and advocacy; program and project management acquisition; strategic planning and policy training, education, and awareness (Brecht, 2020).

The final level of professionalism required for security provision by providing conceptualizing, designing, and building secure systems with clear responsibilities. A specialist could have professional skills that securely involve: the provision of risk management; software development; system architecture; system development; system requirement planning; technology research and development; and testing and evaluation (Brecht, 2020).

PROTECTING SECRETS AND ENSURING INTEGRITY

It is known that robotics is associated with the automation industry and communicates with other devices and equipment. A large amount of data is being exchanged, which raises concerns about its protection. The integrity of the data in the robotic systems is constantly threatened by being modified or exposed. This can be prevented using intrusion detection software, strong encryption, and awareness training programs for users of robotic systems (Lacava et al., 2021). This threat occurs because attackers manipulate robotic system sensors and their cameras and because of issues with the authentication approach or lack of robotic system security design (Lacava et al., 2021). The issue of humans acting socially with robotic systems has also raised concerns about their privacy breaches. It proposed to adopt the privacy-by-design concept to consider privacy concerns in the early design and manufacturing stage. Stealth attacks could occur in robotic systems by changing the output of sensors used, and that could raise concerns about human safety if robots changed their movements (Lacava et al., 2021).

INDUSTRIAL APPROACH FOR CYBERSECURITY IN INDUSTRY 4.0 AND 5.0

The Cybersecurity program's customized design is crucial for organizations in the industry 5.0 era. In this chapter, we propose the usage of the NIST Cybersecurity Framework for evaluating the risk of any security threats. The organization could build a starting point that shows the status of their Cybersecurity essentials, followed by the endpoints which show the targeted goal of implementing the NIST Cybersecurity framework. Digital transformation in the Industry 5.0 Era requires modern technology and standards for Cybersecurity.

For example, the AVEVA company used the NIST framework for securing their services by **partnership approach** with Microsoft Azure and Amazon AWS for their cloud hosting services, and they rely on those reliable companies for protecting the data; the company followed ISO 27001/27017/27018 and AICPA SOC 2 standards (van der Merwe, 2020). The company also followed a **shared responsibility approach** by having the cloud service providers responsible for their service, the company itself responsible for its application, and the customers responsible for

Figure 3. NIST cybersecurity framework core structure
Adapted from (the National Institute of Standards and Technology, 2018).

their configurations and permissions and security countermeasures (van der Merwe, 2020). The company also uses the **separation of environment, access control, and administrative privileges approach**: Segregating the development team from operational ones to protect the customer's data (van der Merwe, 2020). They also used a **data protection and application security approach** by encrypting the data with the latest industrial encryption standards, such as TLS 1.2 (van der Merwe, 2020). Finally, AVEVA has deployed security monitoring and analysis tools to all its environments.

SECURITY STANDARDS

Understating the main principles of Industry 5.0 will assist us in further creating a standard framework utilized to protect the manufacturing Cybersecurity environment. Industry 5.0 will be utilized based on the following main principles: mass customization as per the customer comfort and requirements to move from mass production to mass personalization; cultural collaboration by cross-border collaboration; and customer-centric cyber-physical systems, using an intelligent system which is integrating human intelligence with the machines (Pathak et al., 2019). Finally, green computing maintains a sustainable manufacturing environment.

Knowing the above principles emphasizes upgrading the currently used standards to adapt to the needs of Industry 5.0 Cybersecurity.

For example, the AVEVA company proposed to mitigate the risk of cybersecurity breaches using sound design principles; they have adopted different standards from International Electro-technical Commission (IEC) and International Organization for Standardization (ISO), and American National Standards Institute (ANSI), for their security development lifecycle, standards such as:

- IEC 62443-4-1: Security for industrial automation and control systems;
- ISO 27001: Information security management;
- ISO 27017: Information security controls for cloud services; and
- ISO 27018: Protection of personally identifiable information in public clouds (van der Merwe, 2020).

Figure 4 below shows intelligent manufacturing systems' different layers and associated standards.

Figure 4. The pyramid of intelligent manufacturing systems standards
Adapted from (Lacava et al., 2021)

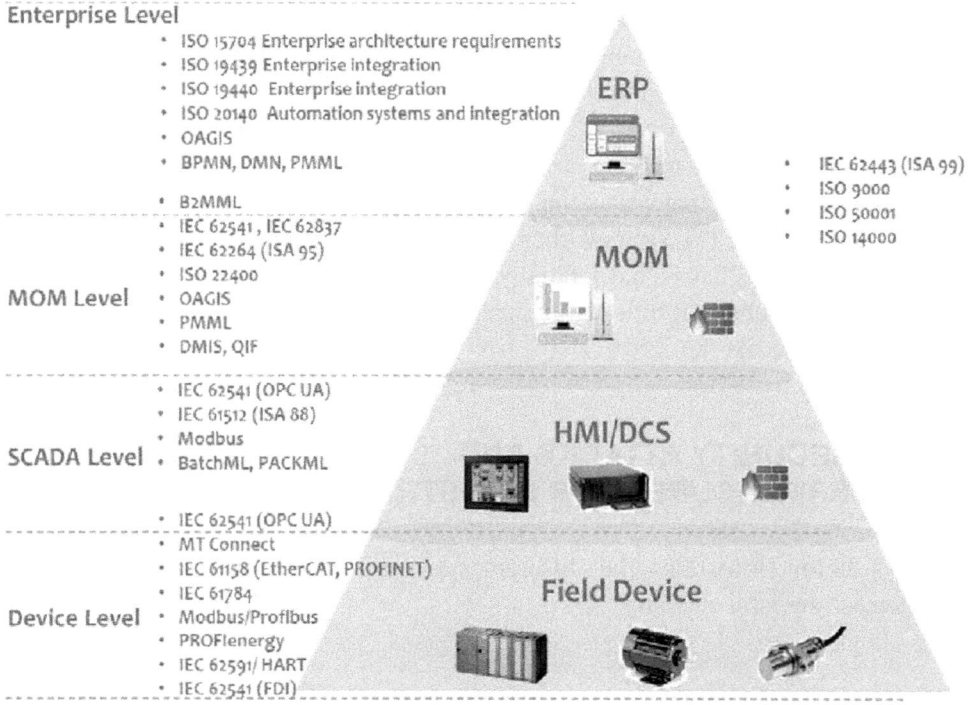

IT AND OT CONCEPTS AND CYBERSECURITY IN THE INDUSTRY 5.0 ERA

(Hakeem, 2019) Industry 5.0 will have the following components for manufacturing personalized medical products, enabling the organization to increase its scale without extra capital investment.

1- Collaborative Robots
2- Additive manufacturing
3- Internet of everything
4- Artificial intelligence
5- Multi-agent systems and technologies
6- Digital Ecosystems
7- Smart Manufacturing
8- Complex adaptive systems

OT refers to operational technology used in manufacturing; it differs from Information Technology (IT). The main differences are: OT is applied as industrial systems, while IT is applied as business systems, and the IT systems cannot stop the work of the OT systems for any reason, such as updating the systems; IT is used for controlling the software applications, while the OT used for manipulating the hardware. The second difference is that we cannot apply the same cybersecurity procedures on IT and OT For example: if the password entered more than once, the IT could lock the user access, while the OT cannot lock access to the industrial operations because of a missing password. The third difference is that OT systems are expected to serve longer than the IT systems, and implement new cybersecurity requirements on the OT systems that may affect the other functionality if it uses older systems (Boo, Patrik, n.d.). Finally, the IT and OT teams should work together for the long-term benefit of the organization and reduce the cost of Cybersecurity attacks (Boo, Patrik, n.d.).

CYBERSECURITY ATTACKS AND COUNTERMEASURES FOR INDUSTRY 5.0

Table 1 below illustrates the different types of attacks and their associated countermeasures for Industry 5.0.

Table 1. Different types of attacks and their associated countermeasures

Cybersecurity attacks	Countermeasures
Social Engineering Attack	Awareness training programs and a strict access control list
Shoulder surfing	Use a password manager, 2-factor authentication, and a privacy cover screen.
Pretexting	Security Awareness Training, Antivirus and Endpoint Security Tools, Penetration Testing, and Security Policy.
Phishing	Do not click on strangers' links, install anti-phishing add-ons, rotate passwords regularly, update the system, install Firewall,
Ransomware	Security Awareness Training, Antivirus and Endpoint Security Tools, Penetration Testing, and Security Policy. Make a backup
Tailgating	Use of Physical Security, Security Awareness Training, and Video Surveillance (Balaban, 2022).
Dumpster	Security Awareness Training, Security Policy, Use of electronic documents always, and use of the shredding facility.
Scareware	Avoid clicking on sudden notifications, update the browsers to the latest version, use the add blockers, Firewalls, and URL filters, avoid suspicious downloads, and avoid clicking on the X button or close the pop-up screens and use the Task Manager using CTRL+ALT+DELETE in Windows systems. Update on the use of Antivirus.
Advanced persistent threat	Security awareness training, Updated Antivirus, and endpoint security tools such as firewalls. Penetration Testing and Security Policy.
Federated identity attack	Use the Automated Lockout after a certain number of failed logins, multi-factor authentication, and Penetration testing.
Denial of service attack	an intrusion detection system (IDS), a strict web server security policy,
Spoofing	Encryption, anti-spoofing, packet filtering, do not take strange phone calls, do not click on strange links in email, and watch the communication for legitimacy concerns.
Man-in-the-middle attack	Encryption, Firewall, VPN, Updated Antivirus, and endpoint security tool.
Back-door attacks	Security awareness training, Updated Antivirus, and endpoint security tools such as firewalls. Penetration Testing and Security Policy
worms	Updated antivirus and endpoint security tools such as firewalls for intrusion detection. Penetration Testing and Security Policy.
Trojans and random access Trojan (RAT) attack	Updated antivirus and endpoint security tools such as firewalls for intrusion detection. Penetration Testing and Security Policy.
Rootkit attacks	Security awareness training, Updated Antivirus, and endpoint security tools such as firewalls. Penetration Testing and Security Policy.
Botnet attack	Updated antivirus and endpoint security tools such as firewalls for intrusion detection. Penetration Testing and Security Policy
Spyware attack	Updated anti-spyware, patched updates, kept backup files, unable to open suspicious files and emails.

continues on following page

Table 1. Continued

Cybersecurity attacks	Countermeasures
Buffering overflow attack	Write secure code, invalidate the stack execution, and check the code using compiler tools and Dynamic run-time check.
Brute force and Dictionary attack	Strong passwords restrict access to authenticated URLs, restrict the number of login attempts, and use CAPTCHAs, and two-factor authentication (Guides, 2019)
Jamming attack	Frequency shifting, frequency hopping
De-Authentication attack	
Eavesdropping attack	Encryption, Security Policy (Kim & Suh, 2021)
distributed denial of service attack DDOS	Intrusion detection systems (IDS), strict web server security policies, security awareness training, antivirus and endpoint security tools, and penetration testing.

REFERENCES

CN Academy. (2022, September). *Cybersecurity Essentials*. Cybersecurity Essentials. https://www.netacad.com/courses/cybersecurity/cybersecurity-essentials

Angel, C. (2022, March 8). *The Biggest Cyber Attacks in the Manufacturing Industry*. Cyber Angel. https://cybelangel.com/the-biggest-cyber-attacks-in-the-manufacturing-industry/

Balaban, D. (2022, February 3). A Step-by-step Guide to Preventing Tailgating Attacks. *CybeReady blog*. https://cybeready.com/a-step-by-step-guide-to-preventing-tailgating-attacks#

Brecht, D. (2020). *What is the NICE cybersecurity workforce framework?* InfoSec. https://resources.infosecinstitute.com/topic/what-is-the-nice-cybersecurity-workforce-framework/

Burton, E., Goldsmith, J., & Mattei, N. (2015). *Teaching AI Ethics Using Science Fiction*. Aaai workshop. *AI and Ethics*.

Demir, K. A., Döven, G., & Sezen, B. (2019). Industry 5.0 and human-robot co-working. *Procedia Computer Science*, *158*, 688–695. doi:10.1016/j.procs.2019.09.104

Ervural, B. C., & Ervural, B. (2018). Overview of cyber security in the Industry 4.0 era. In *Industry 4.0: Managing the digital transformation* (pp. 267–284). Springer. doi:10.1007/978-3-319-57870-5_16

Sucuri Guides. (2019). What is a Brute Force Attack & How to Prevent I? *Sucuri Guides*. https://sucuri.net/guides/what-is-brute-force-attack/

Haleem, A., & Javaid, M. (2019). Industry 5.0 and its applications in orthopedics. *Journal of Clinical Orthopaedics and Trauma*, *10*(4), 807–808. doi:10.1016/j.jcot.2018.12.010 PMID:31316261

Kim, M., & Suh, T. (2021). Eavesdropping Vulnerability and Countermeasure in Infrared Communication for IoT Devices. *Sensors (Basel)*, *21*(24), 8207. doi:10.339021248207 PMID:34960299

Lacava, G., Marotta, A., Martinelli, F., Saracino, A., La Marra, A., Gil-Uriarte, E., & Vilches, V. M. (2021). Cybersecurity Issues in Robotics. *Journal of Wireless Mobile Networks, Ubiquitous Computing and Dependable Applications*, *12*(3), 1–28.

National Institute of Standards and Technology. (2018). Framework for Improving Critical Infrastructure Cybersecurity, Version 1.1 (NIST CSWP 04162018; p. NIST CSWP 04162018). National Institute of Standards and Technology. doi:10.6028/NIST.CSWP.04162018

Pathak, P., Pal, P. R., Shrivastava, M., & Ora, P. (2019). Fifth revolution: Applied AI & human intelligence with cyber-physical systems. *International Journal of Engineering and Advanced Technology*, *8*(3), 23–27.

Siddiqi, M. A., Pak, W., & Siddiqi, M. A. (2022). A Study on the Psychology of Social Engineering-Based Cyberattacks and Existing Countermeasures. *Applied Sciences (Basel, Switzerland)*, *12*(12), 6042. doi:10.3390/app12126042

Stuart, J. (2010). *Artificial Intelligence A Modern Approach* (3rd ed.).

van der Merwe, N. (2020). *AVEVA cloud security: Mitigating cybersecurity risk through good design* [White Paper]. AVEVA. https://www.benelux.avevaselect.com/wp-content/uploads/2022/01/AVEVA-Cloud-Security-Mitigating-Cyber-Security-Risk-Through-Good-Design.pdf

Yaacoub, J.-P. A., Noura, H. N., Salman, O., & Chehab, A. (2021). Robotics cyber security: Vulnerabilities, attacks, countermeasures, and recommendations. *International Journal of Information Security*, 1–44. PMID:33776611

Chapter 4
Industry 5.0 and Cyber Crime Security Threats

Lila Rajabion
SUNY Empire State College, USA

ABSTRACT

Cybersecurity is the act of protecting networks, programs, and systems against various hostile and digital assaults. Subset of a security program, it defends cyberspace from escalating assaults and dangers that result in significant damage to resources like finances, information, and applications. Hackers are increasingly targeting firms in the financial and industrial sectors, particularly for the purpose of stealing sensitive data. As a result, many business leaders are turning to cybersecurity to meet their company's security demands and prevent its precious assets from falling into the wrong hands. When it comes to safeguarding software and hardware against unauthorized access and intrusion, cyber assaults play a critical role. The security measure makes use of several security approaches, such as cybersecurity software, access control systems, antivirus and malware security programs, firewalls, and program upgrades, among all system users.

INTRODUCTION

Cybersecurity is the act of protecting networks, programs, and systems against various hostile and digital assaults. The subset of a security program defends cyberspace from escalating assaults and dangers that significantly damage resources like finance, information, and applications. Hackers are increasingly targeting firms in the financial and industrial sectors, mainly to steal sensitive data (Lezzi, Lazoi, & Corallo, 2018). As a result, many business leaders are turning to cybersecurity to

DOI: 10.4018/978-1-7998-8805-5.ch004

meet their company's security demands and prevent its precious assets from falling into the wrong hands.

When it comes to safeguarding software and hardware against unauthorized access and intrusion, cyber assaults play a critical role. The security measure uses several security approaches, such as cybersecurity software, access control systems, antivirus and malware, firewalls, program upgrades, intrusion detections, and raising security awareness among all system users. As a result of these changes, cybersecurity is now seen as the most robust modern-day defense against digital threats.

Research Objectives

This study's goal was to create new information and reinforce existing knowledge, so it may be shared with other students and the public. Furthermore, to guarantee that students and researchers thoroughly understand the subject, it provides concise explanations of various ideas and procedures and their advantages, hazards, and tactics. The primary aims of this study are to:

- explore the past, present, and potential futures of the cybersecurity sector, identifying which cyberattacks are more widespread; and
- explore the benefits of hyperconnected systems for Cybersecurity for Industry 5.0.

Literature Review

According to Randall and Kroll (2018) study, millions of pieces of personally identifiable information have been stolen from large and small companies across the globe, including the healthcare and government sectors. As a result, several sectors have resorted to cybersecurity to prevent data breaches and theft, secure trade secrets, maintain regulatory compliance, protect sensitive information, and facilitate and enhance corporate operations.

The industrial sector has had to contend with cybersecurity issues for several years, including hackers and malicious intrusions into industrial control systems. These disturbances influence product quality, brand reputation, sales income, market globalization, and the safety of employees.

Industries must modify their business processes and prevent issues influencing their businesses to reduce these damaging interruptions and implement appropriate security procedures. Modern technologies like mobile computing and the Internet of Things have aided strategic decision-making to deter cyber threats, attacks, and vulnerabilities (Lezzi et al., 2018). As a result, these technologies provided industrial control systems with security strategies that enhanced cybersecurity.

Also, integrating many technologies via hyperconnectivity is crucial in preventing cyber-attacks and improving cybersecurity. Hyperconnectivity is vital for safe computing, according to (Dawson et al., 2018), by integrating diverse technologies such as the Internet of Things (IoT), the Internet of Everything (IoE), and the Web of Things (WoT). Using these technologies, employees can connect to other technologies to accomplish the organization's goals, which always prioritize security.

In 2018, Bradley et al. said that the Hyperconnectivity of businesses and organizations throughout the globe had enabled practical security solutions to cyber threats utilizing high-level technologies like the Internet of Things (IoT) to produce detailed insights. As a result of these revelations, cutting-edge approaches and tactics for safeguarding resources from cyberattacks were developed. As a result, security problems in several sectors were solved thanks to working permanently at this site. However, various flaws and dangers to this interconnectedness of technology led to hackability and little ability to alter businesses and guard against cyber-attacks.

Researchers use a variety of techniques, including content analysis, literature review, and other approaches, to gain a better understanding of how various industries use cybersecurity to combat the growing threat of cyber-attacks, safeguard the confidentiality of critical resources, ensure customer satisfaction, achieve a competitive advantage, and protect their company's reputation. Cybersecurity and its hyperconnected systems were shown using qualitative data in this research. Assessment and summation were used to explain the evolution, current, and future trends, common types of cyber-attacks encountered by industries, the hyperconnected systems in cybersecurity used by different industries, the benefits and pitfalls, and strategies to ensure the effectiveness of hyperconnected systems were discussed. The study also employed a digital text format with short and focused paragraphs to boost the audience's reading and clear comprehension of the article. Research papers chronicled the findings and outcomes of the study and were given to the teacher and shared with the rest of the class.

Industry 5.0

Those who collaborate with robots and intelligent machines are referred to as being in 'Industry 5.0.' It is about using cutting-edge technology like the IoT and big data to enable robots to assist people in working more effectively and quickly. Industry 5.0 combines cognitive computing capabilities with human intellect and resourcefulness in collaborative operations as workplace equipment becomes more intelligent and connected. Industry 5.0 has advantages for business, labor, and society. It empowers workers by addressing the employees' changing skills and training demands. It makes business more competitive and aids in attracting top people.

Cybersecurity needs in Industry 5.0

Due to the record high of virtual dangers, cybersecurity is more crucial than ever. Every sector of business must take precautions to guard against cybercrime. Safety should be a top priority in select specific industries. These sectors are the most vulnerable to cyberattacks, which have the potential to cause billions of dollars in damages. This includes a variety of industries, particularly those with substantial industrial and retail operations, that have adopted cybersecurity. In the last several years, many firms have had to deal with significant risks, losing a great deal of revenue and essential resources.

A half-century ago, IT departments in businesses were exposed to a challenging and rewarding job in cybersecurity, among the most vulnerable defenses against cyber-attacks. As technology progresses, so does the complexity of cyber threats, shifting from simple assaults on a single device to current and sophisticated attacks on all devices utilized on industry networks; conventional cybersecurity is unable to fight against these attacks, according to Ismail et al. (2018)

Efficiencies, integrity, secrecy, and access control were all missing from traditional cybersecurity. Due to the absence of firewalls and inferior operating systems, these strategies exposed industry resources to unauthorized access. Several incidents in recent years of malicious activity on industry data, systems, and programs have resulted in businesses downsizing. Many industries felt obligated to fix these difficulties to expand and prosper rapidly. Increased research money helped build cutting-edge security solutions, and scientists and intellectuals spent years improving cybersecurity.

As the use of digital technology in commercial operations expands, cybersecurity is becoming more crucial. As organizations grow increasingly digital, cyberattacks and threats get more sophisticated, making the struggle harder. IT departments of these companies must build new security measures and approaches to tackle cutting-edge hazards. For years, the IT department's battle against cyber-attacks was exclusively a priority. However, after the 2014 Sony hack, cyberattacks were declared an international crime and a vital issue for all businesses worldwide (Takiddin, Ismail, Nabil, Mahmoud, & Serpedin, 2020). Consequently, many sectors realized that a breach caused by a cyber assault causes significant damage to a company's operation, brand and performance, and bottom line.

After these losses and damages, industries classified cybersecurity as a critical business objective, a risk-managing enterprise, and a board-level concern. As a result, cybersecurity solutions were improved (Harkins, 2016). The definition of risks as opportunities for a better understanding of reputational risks from breaches and including all stakeholders in cybersecurity decision-making for better measure formulation were other essential modifications to security rules and compliance requirements. Cyber-attacks on enterprises are rising but do not have to cause financial and resource losses. All sectors must integrate digital technologies, build security strategies that respond to consumer wants, preferences, and experiences, raise cyber threat visibility for more straightforward risk mitigation, and employ intelligence services to enhance privacy and secure security data.

Common Types of Cyberattacks in Industry 5.0

Various cyber threats are employed to assault vital resources in businesses and organizations. However, internal workers may also be engaged in cybercrime by gaining unauthorized access to secured data and systems and stealing company resources, leading to a rise in the leakage and breach of business information. There are three types of attacks, which may be divided into the following categories.

Confidentiality has been attacked: this category includes assaults that steal personal and corporate information such as bank data, transaction history, and mobile phone numbers. Attackers then build fake identities to damage or sell them on the dark Web. Phishing attacks use spam emails to deceive people and workers into giving important information. Various attack tactics cause data breaches, poor customer retention, and high operating costs.

Integrity-attack: this category includes data-leak attacks that undermine industry-critical information. These attacks disclose company secrets, tax information, and other sensitive data. (Kurt, Ogundijo, Li, & Wang, 2018) Discuss attacks in which large amounts of data are disclosed to the public to influence customers and cause firms to lose public trust. This has degraded the industry's reputation and allowed competitors to exploit its strengths.

Intrusion: "Ransomware" prevents authorized users from accessing data, apps, or systems unless a fee is paid. Most ransomware, viruses, and denial-of-service attacks require companies to pay a ransom to recover access to their systems and data and remedy the vulnerability. These attacks destroy resources, harm networks, and disrupt commercial operations.

Industrial Cybersecurity Hyperconnectivity

Hyperconnectivity uses several computer devices and systems to maintain a full-time network connection. Hyperconnected systems, therefore, provide improved communications and corporate connectedness via mobile devices, social media technologies, Operating Systems, etc. IoT, cloud computing, robotics, and IoT cyber-physical systems have advanced Industry 5.0. Dawson et al. (2018) said cybersecurity in the digital industrial revolution depends on hyperconnected systems.

Likewise, corporate data systems were hyperconnected. Hyper-connectivity: IoT, cloud computing, robotics, and IoE cyber-physical systems have been merged to drive Industry 5.0 trends.

Hyper-connectivity: IoT provides networked computers with an internet connection and improves communication. Yuce et al. (2016) said IoT increases industrial security using wireless devices, sensors, and Android. IoT allows data collection, processing, and translation into helpful information and exchange, leading to insights and security decisions. In addition, cybersecurity is improved via IoT maintenance and downtime warnings.

IoT links consumer products to the Web: IoT integrates people, technology, data, process, and objects to transfer information into improved actions that enhance industrial revolution experiences. The IoE decentralizes data and allows rapid device input/output. High-quality data enables business analysis and risk assessment.

Internet-Things (IoT): software that adds objects to the Web. IoT's IoT app layer simplifies development. It simplifies internet connection and communication to increase IoT scalability and flexibility. IoT preserves online information exchange, improves clear, continuous communication, and boosts system adaptability.

Cloud-computing: it gathers, cleanses, processes, stores, and delivers business data. Cloud technologies provide data security and integrity, prevent unauthorized access, allow safe backups, improve data accessibility, and facilitate mobility. In addition, cloud computing safeguards data, enhancing cybersecurity.

AI creates machines that learn, self-correct, and reason like humans: AI uses Machine Learning and business intelligence to transform data into relevant information. This method enhances data-driven predictions.

Robots: robotics can move, react, and sense. These robots do repetitive, high-precision activities like riveting and welding. As a result, robots improve quality, efficiency, and prices. In addition, it boosts security detection and reduces threat susceptibility.

Connectivity is key to improving industrial processes: strengthening user capabilities to facilitate digital transformation, frequent inspection of these systems to decrease attack susceptibility and rapid innovation of technologies; moreover, security awareness must be prioritized to increase their efficacy.

Cybersecurity Hyperconnected Systems' Benefits to Industries

Multiple smart devices to access the Internet and exchange ideas encourage the total transformation of corporate operations such as production, product, and service creation, marketing, communication, risk management, and customer service. According to Jimeno Muñoz (2019), hyperconnected cybersecurity systems benefited industries.

Better information sharing accelerates decision-making and threat response: hyper-connectivity helps risk officers find dangerous system behavior in a catastrophe or attack, notifying the security team and accelerating information flow. This aids decision-making in disasters. This boosts cybersecurity and frontline defense.

Safer: Cloud computing and big data help Hyperconnectivity acquire, clean, process, and analyze data. This information helps managers and analysts generate new ideas and reveal hidden patterns, leading to company development opportunities like industry strengths. In addition, strengths assist in deploying human, physical, and financial resources to prevent threats.

Threat visibility increases: Hyperconnectivity gives managers real-time security information. Disasters and hazards help businesses. Better risk awareness and visibility will lead to early preparedness, mitigation, and management, decreasing hostile actions and ambiguity. Visibility may prevent cyberattacks.

Security policies and legislation: technology and integration advancements assist industry managers in improving operations by offering access and usage controls. This prevents internal attacks and increases cyberspace security.

Better department-industry communication: intelligent technologies and IoT improved employee, customer, and partner communication (Radanliev et al., 2020). These technologies improved business-IT and human-IT connections, boosting IT asset use in business operations and product creation. Unauthorized access, invasions, and DoS attempts were blocked.

Commercialization of cybersecurity: Hyperconnected systems improve business security and efficiency. These solutions may: improve cybersecurity and infrastructure security; enhancing company performance, decreasing operating costs; raising customer delight; preserving a competitive edge, and driving product development. Thus, corporate transformation succeeds, but Hyperconnectivity industries fail.

Future Implications of CyberSecurity

With the advancement of contemporary technology, security has expanded beyond IT to affect corporate operations, cross-industry relationships, vital assets, and future assaults. Cybersecurity has grown better, stronger, quicker, and more intelligent, according to Harel et al. (2017). While Blowers (2015), the growth of cyber-technologies and operations will drive advancements in social media, cyber and physical asset protection, and other disciplines. Cybersecurity's future for people and businesses includes the following:

- mobile and smart technologies enable consumers and companies to access, handle, and manage data. It also enables the creation of essential business applications and platforms incorporating digital security design, requirements, testing, and reviews. As a result, all-important places will be secured;
- cloud computing will boost people's flexibility by letting them access data from anywhere, on any device, at any time. These technologies will secure precious resources and boost cybersecurity awareness. Unknown workers will be victims of cyber-attacks on confidentiality, including social engineering, to steal industrial assets. Farshadkhah et al. (2021) stated that all personnel would require additional security training to protect firms;
- many governments see cyber-attacks as a national tragedy. Hence many states and countries secure crucial national infrastructure and cyberspace. This will boost national security, helping companies safeguard information, servers, devices, and data centers; and
- with more students studying IT to get knowledge and expertise in cyber-attacks, the next generation of cyber specialists will be cultivated. These professionals will create and operate advanced security measures to eliminate cyberattacks on cyberspace and enterprises.

Recommendations

Cybersecurity requires strategic planning and execution to secure vital infrastructure where cyberspace is the primary resource. The report advises the following to boost cybersecurity in all industries:

- Regular files and software backups provide privacy and protection against cyber-attacks. Consider using cloud computing to store data on virtual networks that can be readily accessed and recovered. In addition, information backups assist firms in safeguarding corporate information and documents, ensuring business continuity following catastrophes and assaults;
- Regular updates of all devices, apps, and software to close security flaws. Updated equipment and software increase efficiency, add new features, and resolve security problems that may facilitate attacks. It is essential for all sectors, companies, and people for improved and optimal functioning;
- Encourage workers and consumers to utilize only reputable websites when accessing and supplying sensitive information (e.g., SSN, financial data). Attackers are mimicking legitimate websites to steal crucial information.

Focus on using industry sites and other reputable ones when exchanging and getting information;

- Develop a physical and cyberspace-protective environment. Develop and install current security measures to secure cyberspace, programs, apps, networks, information, authentication technologies, cloud, and IT assets. Define threats and hazards as opportunities to analyze and minimize company risks. This raises risk awareness by identifying, detecting, and mitigating hazards faster. Thus, the risk is managed; and

- All workers should be trained on security dangers and countermeasures. To reduce susceptibility and assaults, skills and knowledge about cyber-attack patterns, detecting systems, and protection approaches must be improved. This will assist the industry in resisting cyber attacks, securing corporate operations, and achieving its objectives.

CONCLUSION

Cybersecurity has become a game-changer in a world of increasing cyber threats and attacks on company resources and infrastructure. Cybersecurity has altered sectors by facilitating prompt risk mitigation, greater company performance, product innovation, and corporate efficiency. Many companies embraced the hyper-connectivity of new technology to improve cybersecurity by merging multiple technologies. These innovations boosted corporate efficiency and productivity by speeding up activities. However, poor leadership, limited capabilities, and low-tech advancements led to problems in hyperconnected networks. Identifying and avoiding these problems is crucial to maximizing industry technology expenditures. All sectors should integrate and accept hyperconnected systems in cybersecurity to effectively reduce and manage risks and business operations for sustainable growth. It is a good business strategy to improve industrial cybersecurity to decrease safety, production, financial, and reputational concerns due to the increasing connectivity of OT systems. Regarding cybersecurity and dependability, executives must take a more active role in aligning OT and IT.

REFERENCES

Blowers, M. (2015). *Evolution of cyber technologies and operations to 2035*. Springer. doi:10.1007/978-3-319-23585-1

Bradley, D. A., Burd, N., Dawson, D., & Loader, A. J. (2018). *Mechatronics: electronics in products and processes*. Routledge. doi:10.1201/9780203747735

Dawson, P., Henderson, M., Ryan, T., Mahoney, P., Boud, D., Phillips, M., & Molloy, E. (2018). Technology and feedback design. *Learning, design, and technology*.

Farshadkhah, S., Van Slyke, C., & Fuller, B. (2021). Onlooker effect and affective responses in information security violation mitigation. *Computers & Security*, *100*, 102082. doi:10.1016/j.cose.2020.102082

Harel, Y., Gal, I. B., & Elovici, Y. (2017). *Cyber security and the role of intelligent systems in addressing its challenges* (Vol. 8). ACM New York.

Harkins, M. W. (2016). *Managing risk and information security: protect to enable*. Springer Nature. doi:10.1007/978-1-4842-1455-8

Ismail, M., Shahin, M., Shaaban, M. F., Serpedin, E., & Qaraqe, K. (2018). Efficient detection of electricity theft cyber attacks in AMI networks. Paper presented at the *2018 IEEE Wireless Communications and Networking Conference (WCNC)*. IEEE. 10.1109/WCNC.2018.8377010

Jimeno Muñoz, J. (2019). Cyber Risks: Liability and Insurance. The Extraordinary Risks in a Hyperconnectivity World. *InDret, 2*.

Kurt, M. N., Ogundijo, O., Li, C., & Wang, X. (2018). Online cyber-attack detection in smart grid: A reinforcement learning approach. *IEEE Transactions on Smart Grid*, *10*(5), 5174–5185. doi:10.1109/TSG.2018.2878570

Lezzi, M., Lazoi, M., & Corallo, A. (2018). Cybersecurity for Industry 4.0 in the current literature: A reference framework. *Computers in Industry*, *103*, 97–110. doi:10.1016/j.compind.2018.09.004

Radanliev, P., De Roure, D., Page, K., Nurse, J. R., Mantilla Montalvo, R., Santos, O., Maddox, L. T., & Burnap, P. (2020). Cyber risk at the edge: Current and future trends on cyber risk analytics and artificial intelligence in the industrial Internet of things and industry 4.0 supply chains. *Cybersecurity*, *3*(1), 1–21. doi:10.118642400-020-00052-8

Randall, K. P., & Kroll, S. (2018). The customer is always right. *ABA Journal*, *104*(8), 30–32.

Takiddin, A., Ismail, M., Nabil, M., Mahmoud, M. M., & Serpedin, E. (2020). Detecting electricity theft cyber-attacks in ami networks using deep vector embeddings. *IEEE Systems Journal*, *15*(3), 4189–4198. doi:10.1109/JSYST.2020.3030238

Yuce, B., Ghalaty, N. F., Santapuri, H., Deshpande, C., Patrick, C., & Schaumont, P. (2016). Software fault resistance is futile: Effective single-glitch attacks. Paper presented at the *2016 Workshop on Fault Diagnosis and Tolerance in Cryptography (FDTC)*. IEEE. 10.1109/FDTC.2016.21

Section 3
Industry 5.0: Utilizing Personalized Design

Chapter 5
Personalised
Measurement Design:
Implementing Industry 5.0

Elspeth McKay
https://orcid.org/0000-0001-7547-9616
Cogniware.com.au, Australia

Mahmoud Numan Bakkar
https://orcid.org/0000-0003-4637-4035
Institute of Applied Technology, Abu Dhabi Vocational Education and Training Institute, UAE

ABSTRACT

The digital economy is now upon us, providing new customer-centric business models fostered during the digital transformation of Industry 5.0. Social networks, mobile applications, data analytics, cloud dependency, and the (IoT) Internet of Things were just vehicles for change. Swept up in this rush for automated business operations is awareness of balancing network complexity and personalizing information measurement systems. This chapter takes a five-pronged approach to situate where the concept of personalized measurement design fits within Industry 5.0. The first section explains how digital transformation has merged multidisciplinary specializations. The second section deals with the preparations for personalized measurement design to enhance the flexibility of online assessment practice. The third section shows how social science knowledge society opens an Industry 5.0 pathway to achieve fully automated human-computer interaction (HCI). The fourth section is about designing flexible online assessments. The final section discusses the next generation of learning analytics.

DOI: 10.4018/978-1-7998-8805-5.ch005

INTRODUCTION

Implementing the design of personalized measurement of human-computer interaction (HCI) for Industry 5.0 involves committed rich data examination by experts aware of how big data is transforming business and society (McKay & Mohamad, 2018). Passing from Industry 4.0 towards 5.0 has meant massive disruption for commerce (Marr, 2016). For instance: this sustainable awareness through improved wireless communications (Alsharif & Nordin, 2017) and finding data behaviour patterns amongst the data messiness (Melendez, 2015; Schöch, 2013). However, few authors agree on what constitutes big data, depending on the philosophical stance taken (McKay & Mohamad, 2018)—keeping such data archived for appropriate retrieval poses major global cyber security issues and crosses ethical boundaries.

This chapter discusses five topics to illustrate where the concept of personalized measurement design fits within Industry 5.0. The first section explains how digital transformation has merged multi-disciplinary specializations, such as the community-rich society created by Derek Powazek (2002), concerning his ideas for artful design for connecting real people in virtual places that are now commonplace. The second section discusses practical preparations for personalizing measurement design to eliminate otherwise subjective decisions leading to incorrect and poor-quality outcomes. The third section shows how the social science knowledge society opens an Industry 5.0 pathway to fully automated IoT human-computer interaction (HCI) connectivity (McKay, 2008). Then the fourth section is about designing flexible online assessment practices, while the final section discusses the next generation of learning analytics.

Digital Transformation Merges Multi-Disciplinary Specializations

Digital transformation is not about technology – it's about change. And it's not a matter of if but a question of when and how. (Weill & Woerner, 2018) p:1

Designing for a community involves three rules (Powazek, 2002): Rule-1: tie content directly to the community and connect the homepages of the content tree and the community tree (Figure 1), where each information system level is interlinked, for instance: through an interlinked tree by connecting the dots (see red line).

Rule-2: bury the Post Button (Figure 2), whereby *"there is a proportional relationship to the distance that the post button is from the front door of the site and the quality of conversation on the site"* (Powazek 2002 p:53). The larger the distance, the better the result is.

Figure 1. Two-tree theory
Source: adapted from Powazek (2002) pp:46-48

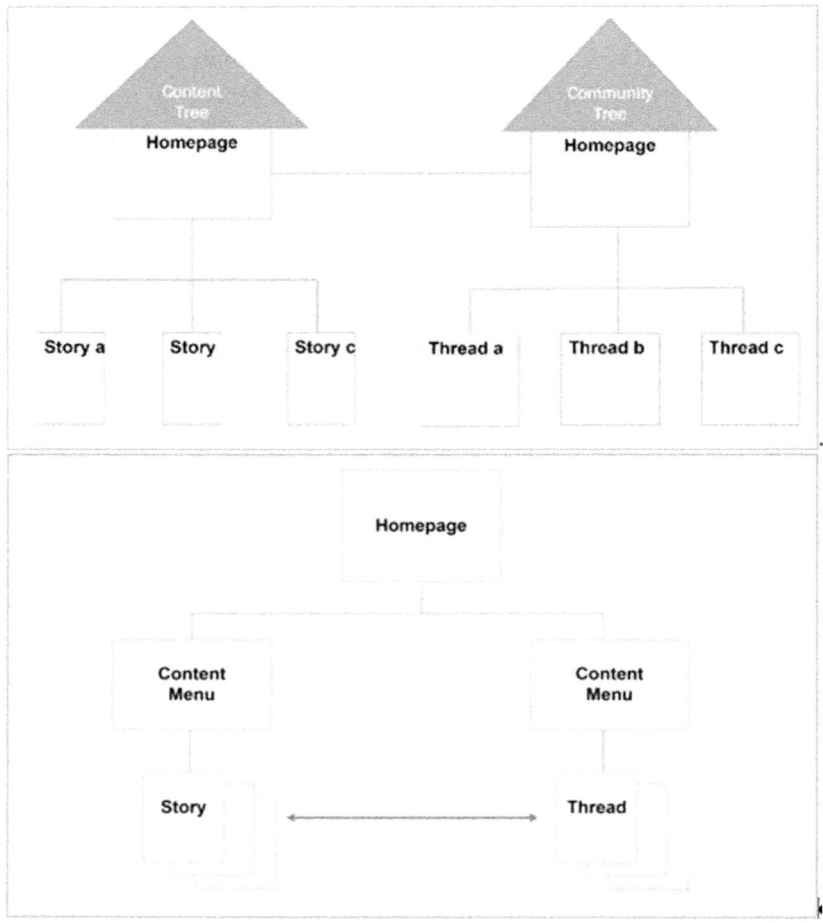

Rule-3: give up control (let the users surprise you). Inviting your users to take the initiative has unseen rewards for the system designer; this is one way to learn things about the user that the designer did not know before, which teaches the designer about how the users view the website. The designer should therefore pay attention to this live feedback to fix the problems as they arise.

Successful business acumen involves understanding how to devise a personalized measurement modelling ruler to deal with a corporate entity's digital disruption in the future (Industry 5.0), according to Weill and Woerner (2018). First, they initiate a two-part self-assessment questionnaire. Part-1 involves short questions concerning the knowledge of the customer(s). Namely, the extent of understanding of the most critical customers, purchase histories, competitors, products for sale, interaction history, business goals (B2B) or personal goals (B2C), and purchase decision-making processes. Then they recommend making a predictive business design (Figure 3). This activity involves understanding the ecosystem necessary to enhance the customers' experience, including a network of partners, suppliers, and other customers. Part-2 of the self-assessment questionnaire involves marking the Part-1 answers. Maximum scores on both the vertical axis and horizontal axis are 100.

Figure 2. Create direct front door access
Source: adapted from Powazek (2002) pp:53-55

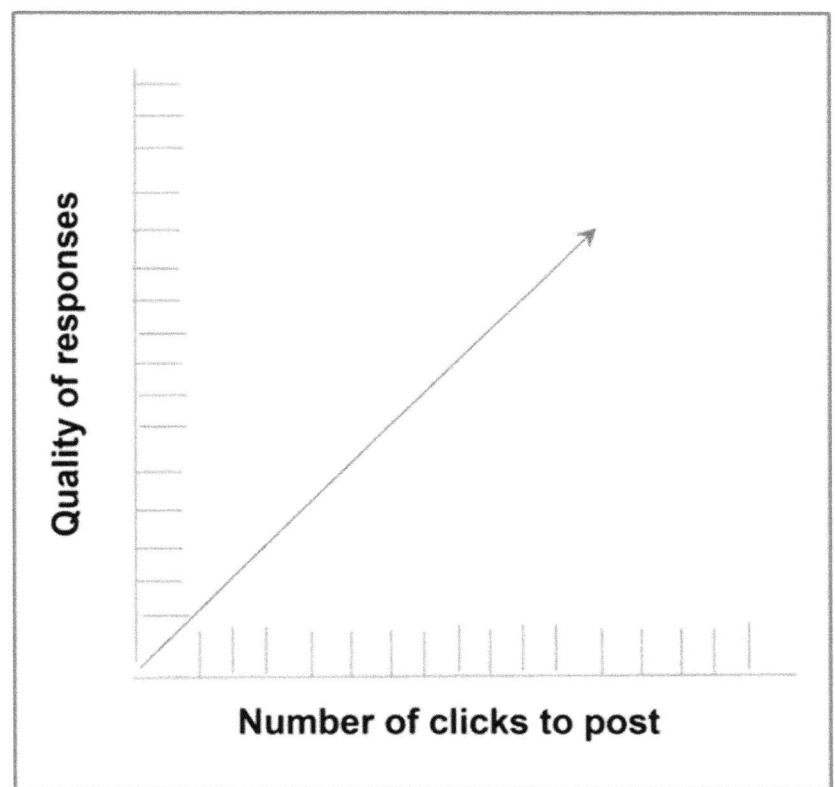

Suppliers have partial knowledge of their end customers
Omnichannels provide customers access to products across multiple channels
Modular producers provide plug-and-play
Ecosystem drivers involve a well-coordinated network of business entities, creating value for all

Figure 3. Weill and Woerner business model
Source: *(Weill & Woerner, 2018) MIT Sloan Centre for Information Systems Research.*

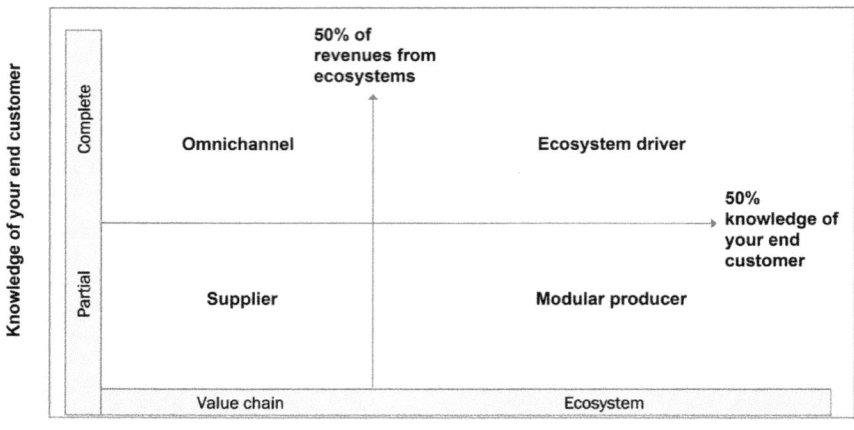

Business Design

Much has been written about trust and trustworthiness, a marketing sector cornerstone and especially critical when designing social capital measurement instruments. However, trust is an abstract human concept (Merrill et al., 1992) that is difficult to measure. For instance: in an experiment reported in The Quarterly Journal of Economics in August 2000, a research project considered measuring two critical components of social capital (Glaeser et al., 2000). This research was a funded project that found attitudinal surveys concerning trust forecast trustworthy behavior better than trusting behavior. Trusting behavior was predicted by past trusting behavior outside of their experiments(Glaeser et al., 2000 p:811) because when

"individuals are closer socially, both trust and trustworthiness rise. Trustworthiness declines when partners are of different races or nationalities. High status individuals are able to elicit more trustworthiness in others."

And so, trust (Mason & Lefrere, 2003) is fundamental to Industry 5.0 business practice. Consequently, understanding the subsequent reliability of the design of the measurement instrumentation is critical. For instance, when the marketing goal is to cultivate communities of practice (Snowden, 2002), the aggregated competencies result primarily from self-organization.

Building trust requires a consistent response from people, the organization and technology and is most effective when operationalized within a context of trust (Figure 4). Therefore, the challenge for Industry 5.0 is to design seamless predictability. Trust cannot be forced, nor can it be designed; it can only be designed for, according to

Mason and Lefrere (2003). Moreover, this means understanding the social capital and cultural aspects, while authorization and authentication are inherent in the technological elements of privacy and security.

Figure 4. Operationalized context of trust
Source: adapted from (Mason & Lefrere, 2003)

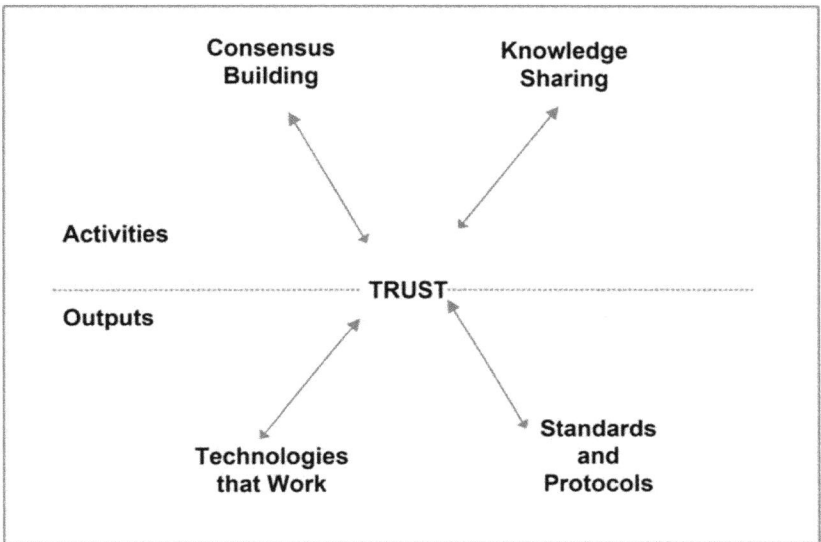

A further challenge facing Industry 5.0 is the effect of the unseen causal differences between each influencing element. Snowden (2002) described this aspect in an idealized Cynefin model to highlight the knowledge elements which flow through three key boundary transactions: the disruption of captured thinking, the creation and stimulation of information communities, and the just-in-time aspect of the informal to formalized knowledge (Figure 5).

Before knowing how to measure the effectiveness of personalized web-based tools, it is necessary to understand the artificial intelligence tool used in the business model. During Industry 4.0, there was a preoccupation within the AI community to capture human behaviors for digital replication (Tecuci et al., 2004) that soon blossomed into the wide range of digital artefacts freely available now through Industry 5.0. Unfortunately, knowledge engineer and subject matter expert roles may be morphed into one person. It would be a retrograde evolutionary system's design issue if this were the case.

Figure 5. Operationalized context of trust
Source: adapted from (Snowden 2002)

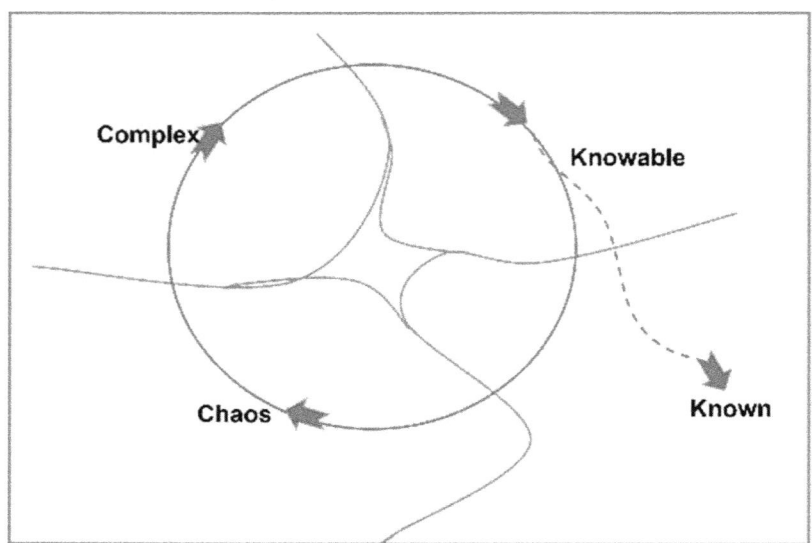

For any particular situation, the knowledge engineer must understand the subject matter problem-solving knowledge to correctly encode it digitally into a database for use by others (Figure 6), provided the problem-solving logic engine has been designed with robust AI rules. Moreover, an SME should examine the solutions generated by an intelligent agent (the information database and the problem-solving application) to identify errors, passing them back to the knowledge engineer for correction (McKay 2016).

Figure 6. Merging IS designer roles
Source: adapted from (Tecuci et al., 2004) and (McKay, 2016)

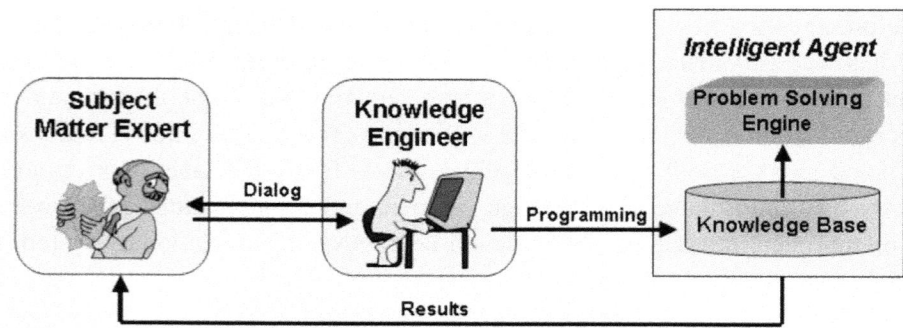

However, as a cautionary Industry 5.0 note that because of the system developer's merging roles (knowledge engineer/subject matter expert), the resulting knowledge acquisition, according to Tecuci et al. (2004), truncates the AI systems development process causes a knowledge acquisition bottleneck (McKay, 2016). For instance: in a three-year funded research project, researchers sought to examine cost-effective online learning systems design using multi-media tools to enhance workplace training with assured predictable outcomes (McKay, 2018). There was a perception from the educational technologists that the most desirable approach to promoting positive eLearning outcomes was to personalize knowledge development through flexible online learning programmes. This project showed IT governance motivated disinterested trainees and energized frustrated management. At the time of the research - multi-disciplined specialists were required to resolve factional dilemmas of governmental IT resource ownership as depicted in Figure 6. The timeliness of this project was to uncover desirable change management issues to improve efficiencies and effectiveness of existing IT training resources.

Relationship Between Personalized Measurement and Industry 5.0

Manufacturing processes, customized measurement and Industry 5.0 are closely related. According to Chen et al. (2010), such decision complexity is required in the Internet age. Therefore, they suggested multiple criteria decision techniques and developed an analytic hierarchy processing (AHP) model. Furthermore, they used comparative baseline systems based on equal weight and rank-based analysis to evaluate the model's outcomes.

According to Lu et al., 2022, industrial human needs can be categorized into five levels: level-1 pertains to security, such as legal rights and protection of human well-being; level-2 pertains to physical and psychological health; level-3 emphasizes human belonging using a different approach such as appreciation, collaboration, and care and on the fourth level; an esteemed approach towards the human-wellbeing needs such as respect. Lastly, level-5 shows the importance of self-actualization reached through personal growth, leadership positions, and self-fulfillment (Lu et al., 2022).

Personalized measurement and Industry 5.0 are connected because both focus on creating products and services that meet the needs and preferences of individuals. Personalized measurement can help achieve the goals of Industry 5.0 by providing the necessary data and information to create highly customized products and services.

The Use of Personalized Measurements in Industry 5.0

Economic and technological interests drive Industry 4.0's goal. The goal is similar to previous revolutions: to increase productivity using innovative technologies (Coronado et al., 2022). However, Industry 5.0 emphasizes the integration of advanced technologies with humans to enhance manufacturing processes. Personalized measurements are crucial in Industry 5.0 as they enable more precise and efficient manufacturing processes that cater to customers' preferences and needs.

Implementing Personalized Measurement Design in Industry 5.0

Coronado et al. (2022) used a systematic literature approach to identify measures, metrics, and quality factors described in the human-robotic interaction (HRI) literature used to evaluate the quality of the HRI. They suggested two models that classified robotics systems' performance-related and human-centered aspects: attitude; acceptance; trust; mental and physical workload; awareness; mental models; and safety. They also proposed seven emergent research areas, which can be applicable in the next years to construct Industry 5.0 applications such as the no-invasive monitoring of human factors, the online analysis of human factors, transparent robotic systems, fluency, privacy in data-driven HRI, benchmarking, and adaptive workload systems.

The need to select the appropriate measurement metrics for evaluating the built Industrial 5.0 applications is crucial in the future. The application builder should be aware of the human factors for facilitating the selection of the metrics needed (Coronado et al., 2022).

In Industry 5.0, personalized measurement design requires a comprehensive approach that combines data analysis, modeling, technology integration, and human expertise. By using personalized measurement designs, companies can enhance customer satisfaction and business performance by developing tailor-made products and services that cater to each customer's specific preferences and requirements.

The next topic discusses the preparation necessary for capturing complex personal characteristics before using Industry 5.0 automation features for personalized measurement design.

Preparing for Personalized Measurement Design

People may use many Industry 5.0 data modelling tools to investigate marketing design effectiveness. For instance: supporting healthcare decision-making for an appropriate choice of the retraining program; closing a project successfully as business executives examine their value-chain management; or building provenance networks from real-world models (Huynh et al., 2018); to the education sector for making decisions more scientifically such as learning-data analytics (Hunt, 2018). Sooner or later, anyone who works with data needs to perform analytics at some point. Consequently, data analytics training courses are widespread and attract analysts and non-analysts alike to learn how to identify their data sets, interpret the resulting metrics and summarise their findings using graphical charts and countless tabulations. Business data analytics involves business-to-consumer software tools. Global digital business enterprises collect and analyze such data concerning their customer base, competitive advantage, and marketing economics to ensure they stay ahead of the pack. For instance: sophisticated marketing recommendation systems draw data from social networking websites to collect user preference data and community interest trending according to demographics, age, and gender.

You can't manage what you don't measure. ~ Peter Drucker

However, managing personalized measurement design involves investigating a complex mix of the knowledge/skill levels and the participants' ability levels. For instance: think of ways to measure wriggly children to standardize age/size clothing labels (Figure 7).

Figure 7. Standardizing clothing designs
Source: 71tCXN1tqUL.jpg ssl-images-amazon.com

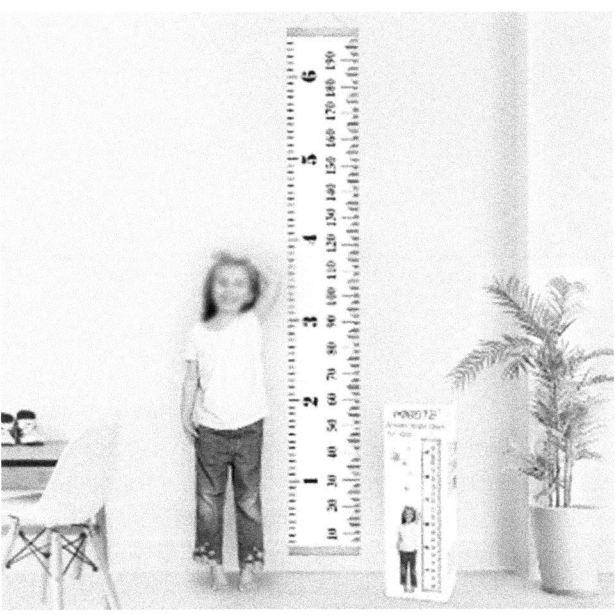

A key feature of the Rasch Measurement Theory (RMT) is the measuring of the participants' scored survey questions (relative to each participant's survey question score) and the behavior of each survey question (relative to the behavior of each survey question), on a common unidimensional RMT scale. The Rasch Unidimensional Measurement Model (RUMM2030) is an application used to check whether data fits the model to provide a reliable estimate of a person's ability profile.

Figure 8 is an example of a RUMM2030 summary statistics table representing the unidimensional Rasch logit-scale location and fit residual statistics (McKay et al., 2022).

Figure 8.
Source: unpublished McKay 2021 RUMM2030 output

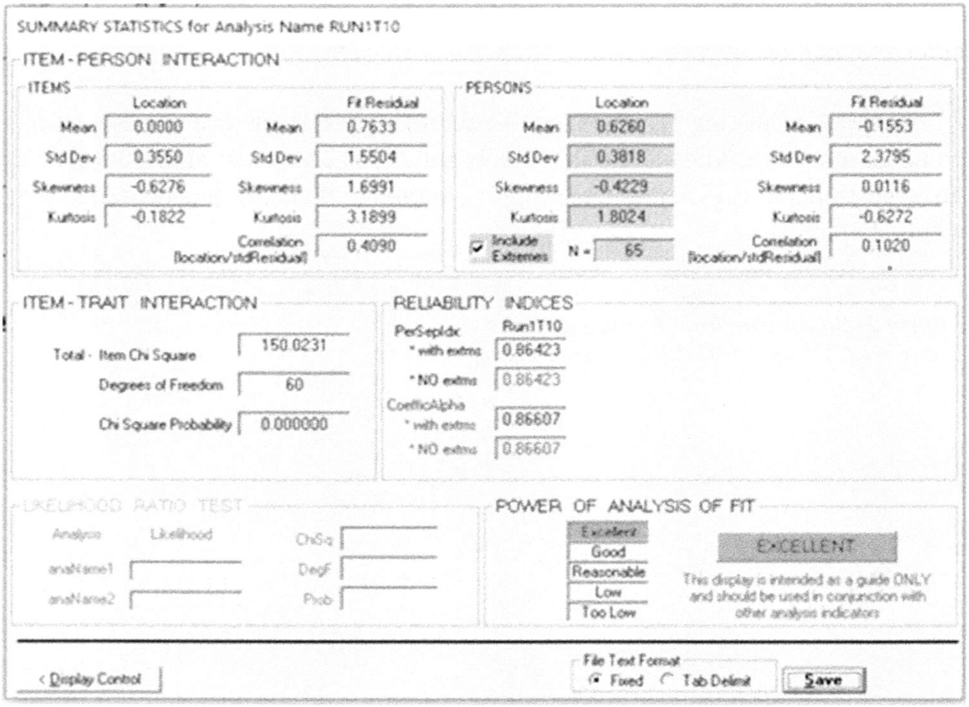

Considering that Industry 5.0 potentially empowers intelligent machines to perform complex actions in collaboration with humans, we pause here to contemplate what must happen first. To make this discussion a little easier for novice statisticians, let us use the proverb that a picture speaks a thousand words; many Industry 5.0 solutions use this approach to communicate the research results. In the case of understanding

Likert scaled survey reactions/attitudes to an online training program, there has to be an investigation of the two dimensions of the scenario to determine the interactive effects of the design of the training program and people's attitudes towards their online training at the end of the session. For instance: the starting point is to examine the results of the online training in terms of the individual participant/person's Likert score relative to each participant's score and the behavior of each survey question relative to each survey question on a common unidimensional scale. The resulting graphical representation of this investigation is shown here as a Person-Item Threshold Distribution below (Figure 9). This data set had 65-participants with a mean logit score of 0.626, a standard deviation of 0.382 and a fit residual mean of -0.155.

Figure 9. Person-Test item distribution outcomes
Source: unpublished McKay 2021 RUMM2030 output

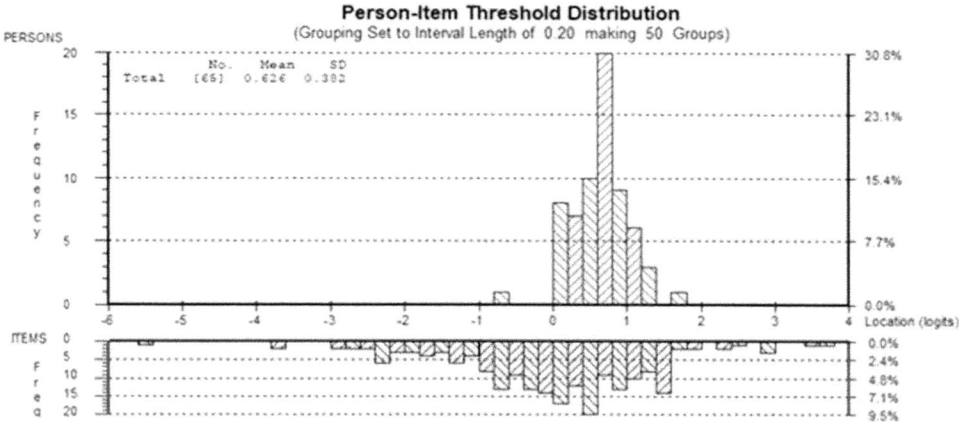

To achieve personalized measurement of a person's suitability for an employment role based on their intellectual capabilities and personality traits, recruitment agencies in the business sector use psychometric assessments, believing that standardized tests provide an objective assessment of the most suitable candidate. These aptitude tests utilized in job interviews include timed questions relating to maths, verbal comprehension and logical (diagrammatic) questions.

Firstly though, the researcher uploads the data file to commence a new RUMM project, specifying: the overall person-item test design; the data format within the data file; and identifying the test-item structures (Figure 10). Once done, RUMM2030 is ready to receive the Display Control Form, involving the conduct of test-of-fit (including ANOVA fit statistics) and saving the Analysis Display (as single/batch files). There are five-main item-person information output categories: 1. item parameter details; 2. test-of-fit details; 3. complete data only; 4. Item characteristics; and 5. further outputs.

Figure 10. RUMM2030 data format form
Source: unpublished McKay 2021 RUMM2030 output

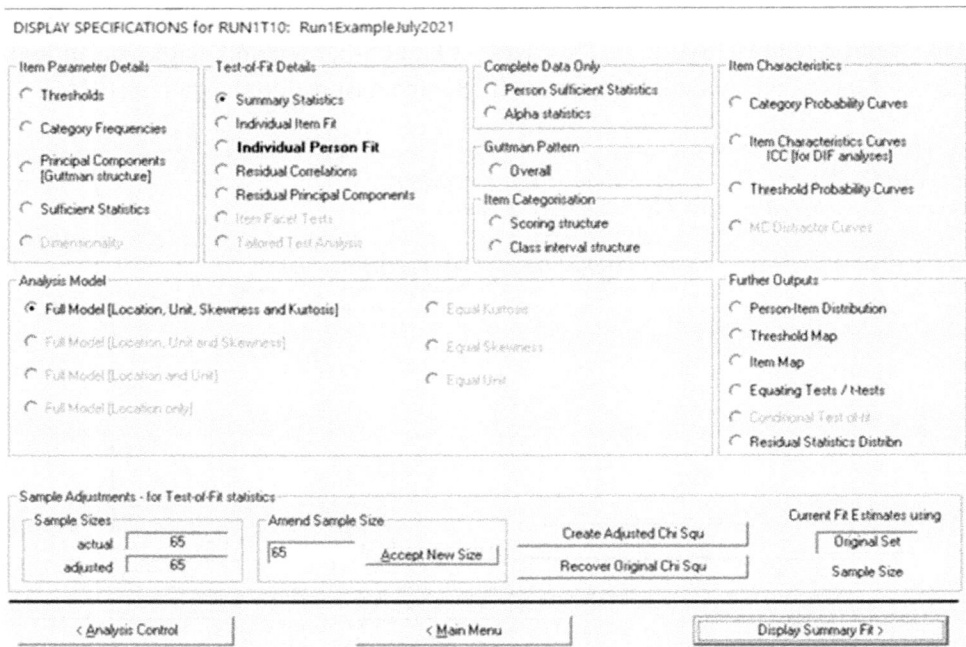

Figure 11 shows another example of research that measured the effectiveness of personalized training concerning the interactive effect of an individual's preferred cognitive information processing and instructional design format (face-to-face, blended, fully computerized) on training performance outcomes (McKay, 2018). The research involved a quasi-experimental 3x3 design, with three independent variables (training mode, instructional training preference and prior domain knowledge (measured as novice, intermediate, and experienced) operationalized through a knowledge/skills screening questionnaire given to participants before undertaking their training session. In addition, the instructional training preferences were categorized as spatial, imagery and verbal using an Object-Spatial Imagery and Verbal Questionnaire (Blazhenkova & Kozhevnikov, 2009).

Figure 11.
Source: McKay 2018 p. 12

Cognitive Media Preference (OSIVQ)	Instructional Treatment		
	Face-to-Face T-1	Blended T-2 (Combination of face-to-face and computerized)	Fully Computerized T-3
Spatial	2	2	6
Imagery	5	5	2
Verbal	1	0	2

Personalized measurement design requires a comprehensive data analysis tool capable of measuring the participants' survey question score (relative to each participant's survey question score) and the behavior of each survey question (relative to the behavior of each survey question) on a common unidimensional Rasch model scale. Figure 12 shows the statistical group fit of 1.39, and the person's ability level of 1.94 in the QUEST kidmap (Adams & Khoo, 1996) revealed through the participant number 2004 performance data it was not a fit of the Rasch Measurement Theory model. However, these performance outcomes were predictive when considering the relationship between the number of easier achieved questions falling in the bottom left quadrant and the lack of harder achieved (top right quadrant). There were eleven questions, which the QUEST estimate considered easy for this participant; there was one harder question not completed with one question completed. This kidmap revealed that this test design could not provide enough evidence of this participant's ability. Instead, it reflected the need for a wider range of questions or increased difficulty in the training module activities.

This section introduced the first chapter topic of digital transformation, merging multi-disciplinary specializations as background, setting the context for Industry 5.0 automating personalized measurement design. The second chapter topic dealt with the importance of preparing for personalized measurement design, with a discussion of the importance of managing the data preparation before using Industry 5.0 automation features. Then, the Rasch Unidimensional Measurement Model (RUMM2023) described a tool that prepares the research data. However, in one sense, this is only the initial data analysis; the next step is to interpret the data to reveal a reliable estimate of a person's ability profile. Once the wholistic picture of personalized ability measurement is known, it is time to illuminate the significance in a broader context, and the next chapter section takes up this discussion.

Figure 12.
Source: McKay 2018 p. 14

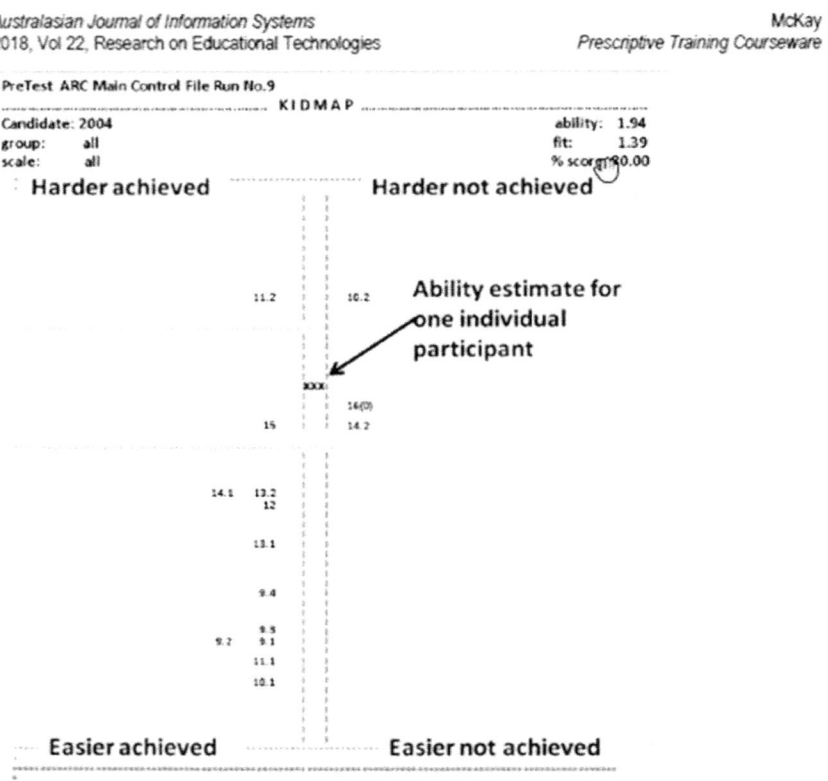

Social Science Knowledge Society Opens a Pathway

The convergence of our roles within society is noticeable through an applied social-contextual lens (Bradley, 2006)

The relationship between HCI and changes in the social and psychological environment (psychological life environment) is central to Industry 5.0. The digital transformation of how people respond to information they receive online follows the earlier predictions of Gunilla Bradley (Professor in Informatics at the Royal Institute of Technology, Sweden (Bradley, 2006). No longer do people isolate their work, professional practice and community citizenship (Figure 13), according to Bradley (2006).

Figure 13. Psychological life environment
Source: adapted from Bradley (2006) and McKay (2016)

Previously, there were niggling anxieties regarding the passage of the so-called fourth industrial revolution (Saniuk, 2022). Industry 4.0 saw many emerging technologies that revamped some people's fears of governments and society, dehumanizing the industry's future, according to Saniuk, Grabowska and Straka (2022); their literature analysis and surveys were considered representative of Polish society. This study unveiled awareness for considering human beings' role in the industry's future development assumptions. For instance: the emerging concerns that Industry 4.0 technologies would become the basis for building the premises of Industry 5.0, thereby blaming the extent of the inevitable business disruption to the digital transformation.

However, the advent of Industry 5 illuminates a social science pathway that self-navigates the complex mixture of automated HCI. A recent social science research study found two essential characteristics of Industry 5.0 (Orlova, 2021). One was the emphasis on personalization not only to provide customers with personalized products but to the personalization of labor relations through the increased HCI traffic (McKay, 2008), connecting the machine-dimension and the human-dimension of HCI through collaborative social communication, loyalty and employees' trust (Orlova, 2021). This study innovated a methodology for investigating corporate human capital assessment and management (CHCM), drawing upon system analysis and synthesis, expert assessments, descriptive statistical analysis and surveys. According to Orlova (2021), human capital (HC) includes natural, physical, economic, human, social and cultural forms of capital to arrive at the following HC typology (Figure 14):

Figure 14. Human capital typology
Source: Orlova(2021) Human Capital Typology p. 7

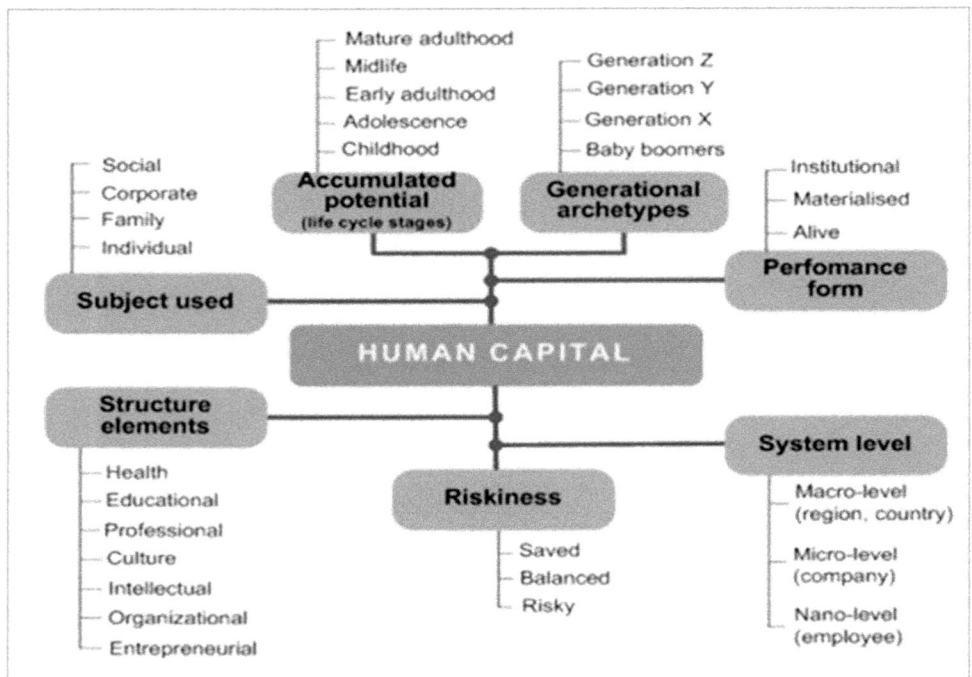

According to Orlova (2021), human capital (HC) comprises the following characteristics: subject used, relating to the social, corporate, family, and individual; performance form, according to the institutional, materialized, or active participation mode; structured elements, for people's lifestyle concerns with health, education, professional, culture, intellectual, organizational, and entrepreneurial; types and components; level of system generational archetypes, accumulated potential and riskiness. The human capital level of these characteristics will change according to the macro/micro/nano levels and properties.

Of course, depending on the system level of individual people in the organization, Orlova (2021) has specific categories of knowledge involvement. For instance: health capital, refers to the lived experience of each employee, therefore reflecting individual health potential; educational capital has complex mixtures of skills and knowledge; professional capital means qualifications and workplace skills and knowledge; intellectual capital refers to levels of creativity, and ability to initiate research and identify new challenges; cultural capital involves a certain mentality, upbringing, ethics and empathic abilities; social capital relates to a person's ability to establish relationships within the organization as well as external social networking during social activities; organizational capital means a person having motivation for engaging in different activities, takes responsibility, has initiative and an ability to set goal to achieve results; and lastly but certainly not the least important aspect of corporate human capital, is the entrepreneurial capital which means an employee's ability to innovate and commercialize results under risky environmental conditions.

Designing for Flexible Online Assessment Practices

Measurement [is] a determination of the magnitude of a quantity, typically on a criterion-referenced test scale or on a continuous numerical scale. Whatever is used to do the measurement is called the measurement instrument. It may be a questionnaire or a test or an eye or a piece of apparatus. In certain contexts, we treat the observer as the instrument needing calibration or validation. Measurement is a common and sometimes large component of standardized evaluations, but a very small part of its logic, that is, of the justification for the evaluative conclusions. (p. 266).

Source: (Scriven, 1991)

There can be no doubt design of online assessment practice has a plethora of meaningful outcomes. This phenomenon was brought about by the just-in-time approach towards fixed deadlines for uploading a fully functional online information system before passing a beta system testing trial. Moreover, much like organizing a camping trip for a social club, it is wise to allow for alternative arrangements for when the weather turns bad or develop an online community course without including the formalized participant/stakeholder feedback. These are lessons learned that form part of the system design's quality control.

Regarding the mounting benefits of Industry 5.0 enhancing people's lives, evidence-based support employment opportunity is typical of the affordances applicable to the healthcare industry that previously required complex programming. For instance: individual placement and support (IPS) specifically address high-risk individuals with persistent and multiple barriers to finding appropriate employment (Kwan et al., 2021).

To fine-tune the data analysis, researchers in Canada used the Rasch Measurement Theory (RMT) to examine the ordering of response option thresholds, fit, the spread of the item locations, residual correlations, person separation index, and stability across time (Barbic, 2014). In addition, this important study examined the emotional vitality of healthcare givers. For instance: Headspace Australia provides a unique understanding of the needs of youth by taking an integrated approach (see Headspace National Youth Mental Health Foundation). The Stepped Care model uses a method of mental healthcare support that is person-centred and supports people across the spectrum of needs. The Stepped Care model aims to ensure that people have streamlined access to the right services for their needs over time and as their needs change. Northern Queensland Primary Health Network (NQPHN) continually examines health outcomes for all residents. Figure 15 depicts their stepped services template, thereby aiding examination of the parts of their model that need adjustment.

Figure 15. Stepped care services model
Source: The Stepped Care model of mental health support (connecttowellbeing.org.au).

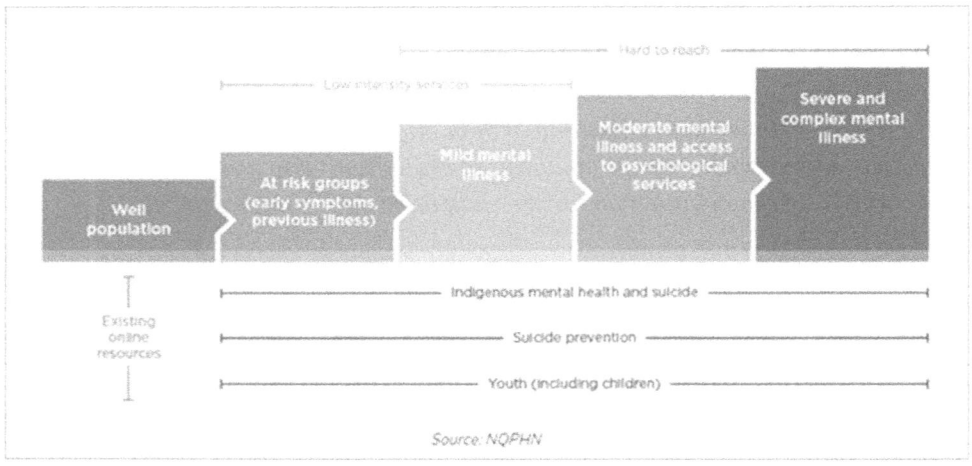

Professor David Williams (Department of Instructional Psychology and Technology, David O. McKay School of Education, Brigham Young University, USA) leads a team of researchers working on cross-cultural evaluation issues, evaluation in schools and universities, evaluation of technology-based teaching and learning, and philosophical, cultural, and moral foundations of evaluation (Williams, 2006) p: 331. This timely book has 19 contributions on related measurement and assessment in evaluation, Web surveying, student feedback, and tests; these chapters illustrate challenges facing online educators in the 21st century. In other words, these educational researchers pushed the boundaries of Industry 4.0 towards Industry 5.0 to provide a collective view. See page xi (Williams, 2006):

- *employ these tools in evaluation systems that support stakeholders;*
- *clarify stakeholder's values and definitions of the evaluands they want to examine;*
- *help them think about the questions they most deeply want answered;*
- *keep the contexts and backgrounds in mind;*
- *seek to adhere to evaluation standards of feasibility, propriety, utility, and accuracy;*
- *and help participants realize that technical issues and methods are only worth it when they assist people in making thoughtful evaluation choices.*

This section discussed a brief outline of the converging roles of human potential for improved psychological life environments; it is time to show how these enhanced automated product services add value to our lives. The next section briefly touches on the possibility that Industry 5.0 personalized measurement techniques will launch the next generation of learning analytics.

Next-Generation Learning Analytics

Integrated smart manufacturing capitalizes on the best features of the machine-dimension of HCI. A theoretical paper (Liu et al., 2019) demonstrated next-generation learning analytics. IoT facilitates cutting-edge technologies such as big data algorithms and programming to enhance intelligent manufacturing and make small to medium enterprises competitive in local markets. This intelligent laundry service represents a fully automated laundry service whereby the HCI applies data (learning) analytics to take advantage of Industry 5.0. Features include: combining input from the city monitoring server to schedule optimal pickup orders from hotel supply partners. Furthermore, reinforced learning algorithms automatically classify orders according to their radio frequency identification (RFID) tags (Liu et al., 2019), which speeds up the laundry service.

While in the education/training sector, smart education and online learning programs are growing in popularity (McKay & Izard, 2013). According to Vladimir Uskov, smart education, smart eLearning, and smart universities potentially transform existing instructional strategies through automated online classroom participation providing students with smart learning spaces (Uskov et al., 2020).

However, there are challenges ahead in the next-generation field of learning analytics. Emerging are worrying ethical developments when automated psychometric testing (unknown to the individual) measures a person's suitability for a role based on their intellectual capabilities and personality traits.

CONCLUSION

The next-generation learning analytics take advantage of Industry 5.0, which benefits many disparate fields of personalized digital measurement design. For instance: measuring the effectiveness of any personalized HCI design for Industry 5.0 involves committed rich data examination by experts aware of how big data transforms business and society (McKay & Mohamad, 2018). No doubt passing from Industry 4.0 to 5.0 has meant massive disruption for commerce.

In the first instance, the digital transformation leading to Industry 5.0 acceptance enabled transdisciplinary merging unheard of during Industry 4.0. Understanding how this changed digital trajectory affects business modelling is essential (Weill & Woerner, 2018). Powazek (2002) contributed to this discourse with a design modelling exemplar involving a rule-based approach to ensure a targeted, personalized measurement modelling ruler. Yet it is the matter of establishing trust that has emerged as an essential and critical component of any Industry 5.0 business practice. Dangers lurk in the amalgamation of systems design roles without enough attention to the intended design pathways remaining on track, thereby leading to reductionist models for effective personalized system design.

Be that as it may, as long as awareness of cultural/socialization characteristics stands up to robust examination, well-designed digital tools have a better chance of designers reflecting high-quality online personalization measurement elements. Rasch measurement theory provides one of the safest data analysis tools to determine how well a personalized digital/online design operates.

Industry 5.0 offers the flexibility of assessment practices for paving a social science knowledge society pathway. Next-generation data/learning analytics leads the way ahead when system developers pay firm conviction and careful attention to the social networking power afforded by Industry 5.0, leading to an exciting future for fully automated personalized online working environments. Moreover, using personalized measurements, Industry 5.0 can improve manufacturing efficiency, quality, and customization, enhancing products and customer experiences.

REFERENCES

Adams, R. J., & Khoo, S.-T. (1996). *QUEST: The Interactive Test Analysis System* (Vol. 1). Australian Council for Educational Research. [Software User Manual]

Alsharif, M., & Nordin, R. (2017). Evolution towards fifth generation (5G) wireless networks: Current trends and challenges in the deployment of millimetre wave, massive MIMO, and small cells. *Telecommunication Systems*, *64*(4), 617–637.

Barbic, S. B., Susan & Mayo, Nancy. (2014). Emotional vitality in caregivers: Application of Rasch Measurement Theory with secondary data to development and test a new measure. *Clinical Rehabilitation*, 29, 29. PMID:25246610

Blazhenkova, O., & Kozhevnikov, M. (2009). The New Object-Spatial-Verbal Cognitive Style Model: Theory and measurement. *Applied Cognitive Psychology*, 23(5), 638–663. doi:10.1002/acp.1473

Bradley, G. (2006). *Social and Community Informatics: Humans on the Net*. Routledge.

Chen, D.-N., Hu, P. J.-H., Kuo, Y.-R., & Liang, T.-P. (2010). A Web-based personalized recommendation system for mobile phone selection: Design, implementation, and evaluation. *Expert Systems with Applications*, 37(12), 8201–8210.

Coronado, E., Kiyokawa, T., Ricardez, G. A. G., Ramirez-Alpizar, I. G., Venture, G., & Yamanobe, N. (2022). Evaluating quality in human-robot interaction: A systematic search and classification of performance and human-centered factors, measures and metrics towards an industry 5.0. *Journal of Manufacturing Systems*, 63, 392–410. doi:10.1016/j.jmsy.2022.04.007

Glaeser, E. L., Laibson, D. I., Scheinkman, J. A., & Soutter, C. L. (2000). Measuring Trust. *The Quarterly Journal of Economics*, 115(3), 811–846. doi:10.1162/003355300554926

Hunt, R. s. (2018). Learning Data Analytics Carpenteria, CA, linkedin.com.

Huynh, T. D., Ebden, M., Fischer, J., Roberts, S., & Moreau, L. (2018). Provenance Network Analytics: An approach to data analytics using data provenance. *Data Mining and Knowledge Discovery*, 32(3), 708–735. doi:10.100710618-017-0549-3

Kwan, A., Morris, J., & Barbic, S. (2021). 811Protocol: A mixed methods evaluation of an IPS program to increase employment and well-being for people with long-term experience of complex barriers in Vancouver's downtown and DTES. *PLoS One*, 16(12), e0261415. PMID:34914771

Liu, C., Li, H., Tang, Y., Lin, D., & Liu, J. (2019). Next generation integrated smart manufacturing based on big data analytics, reinforced learning, and optimal routes planning methods. *International Journal of Computer Integrated Manufacturing*, 32(9), 820–831. doi:10.1080/0951192X.2019.1636412

Lu, Y., Zheng, H., Chand, S., Xia, W., Liu, Z., Xu, X., Wang, L., Qin, Z., & Bao, J. (2022). Outlook on human-centric manufacturing towards Industry 5.0. *Journal of Manufacturing Systems*, 62, 612–627. doi:10.1016/j.jmsy.2022.02.001

Marr, B. (2016). *Big Data in Practice: How 45 Successful Companies Used Big Data Analytics to Deliver Extraordinary Results*. John Wiley & Sons, Ltd. doi:10.1002/9781119278825

Mason, J., & Lefrere, P. (2003). Trust, Collaboration, e-Learning and Organisational Transformation. *International Journal of Training and Development*, 7(4), 259–270.

McKay, E. (2008). The Human-Dimensions of Human-Computer Interaction: Balancing the HCI Equation (1 ed., Vol. 3). IOS Press.

McKay, E. (2016). Gearing up the knowledge engineers: Experience design through effective human-computer interaction (HCI). In A. Lugmayr & C. Dal-Zotto (Eds.), *Media Convergence Handbook* (Vol. 2, pp. 283–307). Springer-Verlag. doi:10.1007/978-3-642-54487-3_15

McKay, E. (2018). Prescriptive Training Courseware: IS-Design Methodology. *AJIS. Australasian Journal of Information Systems*, 22. https://doi.org/http://dx.doi.org/10.3127/ajis.v22i0.1675

McKay, E., Asquith, K., Smyrnova-Trybulska, E., Porczyńska-Ciszewska, A., & Kopczyński, T. (2022). Data Evaluation of Happiness Scale Online Study: A Rasch Measurement Analysis. In E. McKay (Ed.), *Manage Your Own Learning Analytics: Implement a Rasch Modelling Approach* (Vol. 261, pp. 73–112). Springer. doi.org/10.1007/978-3-030-86316-6

McKay, E., & Izard, J. (2013). Seamless Web-Mediated Training Courseware Design Model: Innovating Adaptive Educational-Learning Systems. In A. P. Ayala (Ed.), *Intelligent and Adaptive Educational-Learning Systems: Achievements and Trends, ISSN:2190-3018 (Vol. 17*, pp. 417-442). Springer Berlin Heidelberg.

McKay, E., & Mohamad, M. (2018). Big Data Management Skills: Accurate Measurement. *Research and Practice in Technology Enhanced Learning*, 13(5). doi:10.118641039-018-0071-2 PMID:30595736

Merrill, M. D., Tennyson, R. D., & Posey, L. O. (1992). *Teaching Concepts: An instructional design guide* (2nd ed.). Educational Technology Publications.

Orlova, E. V. (2021). Design of Personal Trajectories for Employees' Professional Development in the Knowledge Society under Industry 5.0. *Social Sciences*, 10(427). doi:10.3390ocsci10110427

Powazek, D. M. (2002). *Design for Community: The Art of Connecting Real People in Virtual Places*. New Riders.

Saniuk, S. G. S., & Straka, M. (2022). Identification of Social and Economic Expectations: Contextual Reasons for the Transformation Process of Industry 4.0 into the Industry 5.0 Concept. *Sustainability*, *14*(1391). doi:10.3390u14031391

Scriven, M. (1991). *Evaluation Thesaurus* (4th ed.). Sage.

Tecuci, G., Boicu, M., Ayers, C., & Cammons, D. (2004). Cognitive assistants for analysts. In *The Proceedings of the 4th Annual Analysis & Productions's Analysis Congerence: The future of Analysis.* NSA. http://lac.gmu.edu/publications/2007/ TecuciG_Cognitive_Assistants.pdf

Uskov, V. L., Bakken, J. P., Gayke, K., Fatima, J., Galloway, B., Ganapathi, K. S., & Jose, D. (2020). *Smart Learning Analytics: Student Academic Performance Data Representation, Processing and Prediction.* Springer Singapore. doi:10.1007/978-981-15-5584-8_1

Weill, P., & Woerner, S. L. (2018). *What's Your Digital Business Model? Six questions to help you build the next-generation enterprise.* Harvard Business Review Press.

Williams, D., Hricko, M., & Howell, S. L. (Eds.). (2006). *Online Assessment, Measurement and Evaluation: Emerging Practices. IGI Global.* IGI Global. doi:10.4018/978-1-59140-747-8

Chapter 6
Personalized Product Recommendation and User Satisfaction:
Reference to Industry 5.0

Priyadarsini Patnaik
Birla Global University, India

Parameswar Nayak
Birla Global University, India

Siddharth Misra
Birla Global University, India

ABSTRACT

The transition from Industry 4.0 to Industry 5.0 started when personalization options became available to customers. This revolution aims to bring back the human touch with the convergence of advanced technology towards a degree of personalization to meet the demand of the customers. In the era of Industry 5.0, consumers want to differentiate themselves as unique, and personalized products allow them to express themselves as individuals. This has prompted personal recommendations to become more popular. Despite the increasing popularity of personalized recommendations, little research has been conducted on the impact of these recommendations on user satisfaction. As a result, an online survey was conducted to test the relationships between personalized product recommendations and user satisfaction and proposed a conceptual model. The findings of the study indicated a positive association between personalized product recommendations and consumer satisfaction and highlighted several managerial and practical implications that academics and retailers may find useful.

DOI: 10.4018/978-1-7998-8805-5.ch006

INTRODUCTION

Innovation, personalization, and customization are the three aspects where Industry 4.0 falls short (Doyle-Kent & Kopacek, 2019) since, in today's market, mass customization fuelled by Industry 4.0 technologies is inadequate to satisfy highly personalized customer demands (Ostergaard, 2018). Hence, Industry 5.0 initiated the fundamental foundations of personalization (Iyengar et al., 2022) and allowed for product personalization with higher human intelligence engagement (Akundi et al., 2022). The transition from mass production primarily drives mass customization, where traditional marketing is transformed to create a super-empowered customer with innovative digital information technologies. Consumer demand for mass customization is the primary force behind Industry 5.0. Since of this insight, consumers are willing to pay a premium for Industry 5.0 products because they allow them to satisfy their self-expression needs. In a nutshell, the goal of the Industry 5.0 paradigm is people-focused and robust, providing an option for customers to select products tailored to their specific preferences and requirements. This revolution aimed at bringing back the human touch with the convergence of advanced technology towards a degree of personalization to meet the customers' demand. This technological advancement makes it possible to achieve high performance and allows a high personalization level to fulfill demands specific to individual customers. Researchers have identified several enablers for Industry 5.0, including AI, IoT, and Big data, which help in enterprise digitization and innovation (Maddikunta et al., 2022). This human-machine interaction can be used in various ways to influence consumer behavior. In this technology era, the rapid proliferation of internet information and services makes it challenging for users to make clear decisions regarding e-shopping. Hence, recommendation systems are designed to alleviate the problem of information overload by recommending products based on a user's interests. As a result, many online recommendation systems have been created to assess users' preferences for product qualities and help them make more informed purchases (Ghasemaghaei et al., 2019). Online Product Recommendations are the recommendations generated based on the user's preferences (Bathla, 2017). Recommendation systems use a variety of strategies, including content-based, collaborative filtering, and trust-based recommendations. A collaborative filtering algorithm analyses the user's browsing history and current activity to generate personalized recommendations to forecast their possible future behavior (Wu, 2021). To predict future purchases, OPR analyses a customer's recent purchase, purchase frequency, and transaction amount of past purchases (Nassar et al., 2020). Hence, e-commerce websites utilize product recommendation algorithms to help users discover new products more quickly and easily.

Traditional algorithms for making recommendations take into account past user activity and create suggestions based on their possible shared preferences (Hui et al., 2022; Fu et al., 2020). User-item interactions, on the other hand, tend to be highly sparse, and new user's lack of relevant information can cause cold start issues if recommendations are based solely on previous behavior (Wen et al., 2022). Hence, personalized recommender systems hold great promise for addressing the issues of information overload, overstimulation, and information disorientation (Wang et al., 2022). Nevertheless, customers cannot get prompt and accurate recommendations from traditional recommendation algorithms, which leads to low recommendation efficiency. As a result, users were given the option of a personalized recommendation, and Big data is used to create these personalized recommendations, which include products, services, and information tailored to each customer (Subramanyan, 2014). With the help of personalized content, businesses can develop personalized online consumer experiences tuned explicitly to the wants and needs of every one of their customers. Hence, much relevant content can be delivered via this artificial intelligence (AI) based hyper-personalization, which takes personalized marketing a step further. With the help of these technologies, customers can now design their environments depending on their preferences, interests, and beliefs. Companies can also use this approach to tailor information to their customer's specific needs. Hence, hyper-customization used by marketers to provide clients with customized information. A recent development in artificial intelligence-enabled microtargeting marketing, also known as personalized content recommendation algorithms (André et al., 2018). Personalized recommendation recommends new content to users based on their unique preferences (Covington et al., 2016; Gomez-Uribe & Hunt, 2015). It comprises 79% of AI inference (Gupta et al., 2020). According to A. Eisenman et al. (2018) and Naumov (2019) highly accurate personalized recommendations can be accomplished by user preferences and their past interactions, whereas it is achieved using deep neural networks (Gupta et al., 2020). Research has shown that personalized recommendations lead to positive responses toward the brand and enhance behavioral intentions (Maslowska et al., 2016). Hence, Artificial Intelligence (AI) is helping firms create more tailored consumer experiences to forecast what customers want or need and which customers will purchase the item. So, AI and its applications will eventually be accepted by the company, and this trend will continue. Personalization has grown increasingly important as the world gets increasingly digital. Artificial Intelligence (AI) constantly improves the customer experience in traditional marketing (AI).

Study Objective

Hyper-personalization has arisen as a step beyond personalization in recent years. The importance of customer-centric services in the realm of electronic commerce is growing for this same reason. This industrial revolution 5.0 is making customers happier by improving customer satisfaction with the help of personalized products. Hence, businesses use personal product recommendations (PPRs) to provide their consumers with more personalized experiences, anticipating their needs, and wants and predicting the number of customers who will purchase the product in the future. Personalized marketing is taken to a new level when these predictions are made based on a customer's browsing history, spending habits, and personal preferences. As a result, the adoption of PPRs by the company has increased. However, customers' use of these recommender systems has yet to be studied in terms of user satisfaction, despite industry and academia giving them considerable attention. Also, whether customers will use and be satisfied with PPRs is critical. To a large extent, the recommender system's success hinges on its ability to deliver products well-suited to individual preferences. However, if it always suggests the same thing, customers will not be as satisfied, no matter how accurate the recommender system is (Aggarwal, 2016). Thus, although research on recommender systems has primarily focused on boosting the model's performance, it is just as crucial to ensuring that users are satisfied with the system. However, when evaluating customer satisfaction, the PPR literature receives scant attention (Kim et al., 2021; Sheng et al., 2014). Finding empirical research that investigates the connection between personalized offerings in e-commerce and consumer satisfaction is difficult. Therefore, this research aims to determine how personalized recommender systems in electronic commerce impact consumer satisfaction and how customers feel about personalized recommender systems.

Literature Review

Personalization in Industry 5.0 Era

Cartwright presented a detailed overview of Industry 5.0 in 2018, where he predicted collaboration between man-machines under Industry 5.0. According to him, the most prominent is the crucial feature of Industry 5.0, where users can directly access sensor data to design and manufacture personalized products in real-time, facilitating human-machine interaction and cooperation (Di Nardo et al., 2021). Personalized experiences and bundles of products/services are what customers expect. Customers want to stand out, be recognized for their distinctive tastes, and display their individuality through their purchases. As a result, products and services become more personal with Industry 5.0 (Ostergaard, 2018). To meet the personalized demands of the consumer, Industry 5.0 aims to improve collaboration between humans and manufacturing systems (Javaid & Haleem, 2020). Industry 5.0 has gained popularity due to today's consumers' rapidly increasing individualized requirements. As a result of mass personalization, artificial intelligence has been integrated into human life to boost human capabilities (Martynov et al., 2019). As a result, human and machine workers will be better synchronized through the fifth industrial revolution to accomplish faster and more efficiently. Furthermore, thanks to the flexibility of human-computer systems, it can meet and even exceed a customer's demands (Ozdemir & Hekim, 2018).

Personalized Product Recommendation System

Consumers today have easy access to digital marketplaces, great product choice, immense product-related information with great convenience, an enormous array of products, and lots of information about their chosen products. However, since human information processing is limited in cognitive capacity, finding products that meet consumers' interests is critical. So, technological tools are being used by many online stores to assist in product selection and search. Personalized Product Recommendation agents (PPRAs) are examples of such tools. PRAs are personalized web technology that provides recommendations to individual consumers based on their preferences and needs regarding products related to their buying (Xiao & Benbasat, 2014, 2015, 2018).

These product recommendations are essential to Artificial Intelligence (AI) to provide users with better and more helpful service. Due to the vast amount of data, users need to get fast and accurate recommendations from traditional algorithms, so personalized recommendations were suggested (Cui et al., 2020). Nevertheless, consumers are empowered when PPRAs offer true personalization based on their preferences and what their customers want (Xiao & Benbasat, 2018). So, online retailers use web-based recommendations to assist consumers with searching and selecting products while shopping (PPRAs). The recommendation of a product is based on the user generated content (UGC), like user reviews, opinions, experiences, browsing history, and recommendations of products and services (Zhang et al., 2019). Researchers have looked at recommendation behavior in the context of e-commerce sites and found that consumers' online choices are influenced by information from recommendations (Liu et al., 2016).

The recommendation system assists users in making purchases by predicting what items they may wish to purchase in the future based on the symmetry between their behavior and the characteristics of their actions. A recommendation system is a component of machine learning algorithms for recommending relevant suggestions to the user (Shahbazi & Byun, 2019). Recommendation systems can be classified as content-based and collaborative filtering. With content-based filters, the user can see items similar to the items they have previously purchased, and collaborative filters present similar items to the item selected by the user (Changchien et al., 2004; Balabanovic & Shoham, 1997; Cunningham et al., 2001; Cho et al., 2002). Compared to other approaches, collaborative filtering is a critical element of the modern paradigm of recommendation systems due to its ability to capture the interest of a user effectively (Shahbazi et al., 2020).

User Satisfaction

Oliver (1997) defined customer satisfaction as the happiness of a single emotional state or the fulfillment state at a pleasant level. According to Lee (2007), customer satisfaction is the sum of customer emotions following a product or service's initial experience or evaluation. Two perspectives can be taken on customer satisfaction. First, customer satisfaction generally refers to the difference between the customer's expectations and the service they receive (Zeithaml et al., 1993). To determine customer satisfaction, it is essential to evaluate service function within a short period (Grönroos, 1994). However, Bendapudi and Berry (1997) stated that customers are satisfied with a specific service provider when they have a positive experience with them that takes a broader perspective.

Researchers and experts have created a variety of studies and models to study customer satisfaction, categorized as macro- or micro-models. Several countries have developed national customer satisfaction indexes to measure customer satisfaction with products and services since the 1990s. For example, Fornell (Fornell,1992) proposed the customer satisfaction index (CSI) in 1989 by assessing both customer expectations and perceptions after purchasing. In addition, Parasuram et al. created a SERVUAL scale to evaluate service quality (Parasuraman et al.,1993). He says, "five factors have been identified as contributing to service quality: reliability, responsiveness, assurance, empathy, and tangibility."

As online services continue to grow, user satisfaction is more important than ever. (Dianat et al., 2019) In addition to influencing customer behavior such as loyalty and trust, user satisfaction is also critically important since it influences the intentions of customers to purchase (Dash et al., 2021; Dhingra et al., 2020; de Morais Watanabe et al., 2019; Hossain et al., 2018). It also contributes significantly to profitability (Eklof et al.,2020; Salameh et al., 2020). Researchers studied user satisfaction from various perspectives (Elavarasan et al., 2021; Kwon et al., 2020; Santa et al., 2019). "A model of user satisfaction developed by Doll et al. (1994) consists of five system attributes to consider, i.e., content, format, accuracy, timeliness, and ease of use." Also, user satisfaction is predicted by the content of commercial websites and their search capabilities (Zviran et al., 2006). However, customers will be more likely to buy from OPRs in the future if they are satisfied with the results of their expectations being met and have confidence and trust in OPR.

Customer satisfaction plays a vital role in the e-commerce environment in attracting and retaining customers (Tandon et al., 2017). In addition, a customer's satisfaction with a product or service is an essential determinant of whether they continue or discontinue using it (Chen et al., 2020; Chung & Shin, 2010). Therefore, one of the keys to the long-term success of an online store is customer satisfaction, which can lead to improved customer retention (Chen et al., 2012) and repurchase intention (Yiu et al., 2007).

Product Recommendation and Satisfaction

Customers perceive satisfaction while receiving personalized recommendations online (Jiang et al., 2010). Generally, if clients buy the products recommended by a recommendation system, that is considered a success. However, a purchase in and of itself does not ensure satisfaction; recommendation systems need to maximize customer satisfaction when recommending products or services to users. The recommendation system recommends a product to a customer based on their characteristics to predict that a high level of satisfaction will be experienced. It is possible that the recommended product(s) were purchased, but the recommendation system failed to satisfy the customer, which is its ultimate objective. Hence, accepting a recommendation by a customer does not necessarily equate to its success. For recommendations to be effective, customer needs must be met while the customer wants must also be satisfied. A recommendation system must, therefore, only recommend a product if it is likely to have a high satisfaction rating. Online product recommendations and quantified also value this customer satisfaction. In this study, customer satisfaction is calculated as the degree to which customers are satisfied by purchasing the recommended products. Here, consumer satisfaction includes the quality and performance of the recommended product and how much the customer meets the expectations from the recommended product. The degree to which users are satisfied with their experience plays a crucial role in evaluating the effectiveness of OPR (Hou, 2012). Also, consumers make purchases by evaluating the quality (Parasuraman et al., 1993), accuracy (Doll et al., 1994), and price of any product or service (Walia et al., 2016; Karmarkar et al., 2015; Kim et al., 2012) while shopping on any commercial websites. In this study, quality refers to the recommended quality of the personalized product recommender system, accuracy refers to recommendation accuracy, and price refers to the price recommended during product recommendation. Hence, this study considered these three characteristics of personalized product recommender systems and examined how these characteristics impact customer satisfaction.

Recommend Accuracy

Recommendation accuracy is high while keeping consumers' efforts low (Pfeiffer et al., 2013). Herlocker (2004) defines accuracy as the fit between predicted and actual customer preferences. A study by Liang et al. (2007) indicates that user satisfaction can be increased based on the accuracy of the recommendation. Therefore, if the user is provided with recommendations that suit their preferences, they will be more inclined to accept them. Accuracy is an essential characteristic in recommendation systems by keeping user prediction ability as the most crucial characteristic. The accuracy of predicting the relevant products to be recommended is most likely to be purchased by the customers (Shahbazi et al., 2020). Thus, many previous studies have looked at ways of improving the predictability of the recommendation system. Hence, the user can evaluate the recommendation system using the accuracy of the recommendation by evaluating its usefulness.

Price Fairness

Price has been documented as a critical determinant of purchasing decisions by consumers (Walia et al., 2016). Comparison shopping is a standard method for consumers to choose one item over another, generally selecting the less expensive option. However, consumers agree upon an alternative model that considers the price a determinant of quality. Therefore, they may consider high-value products to reduce the quality risks associated with e-shopping (Palma et al., 2016). On the other hand, any product purchased at an affordable price, which is neither too cheap nor too expensive, will tend to attract consumers considering budget and quality. Thus, a higher expectation of price acceptability will result in more sales (Huang et al., 2019). Holbrook (1996) said that if a vendor recommends a product without considering price response, the impact will be small; however, if the vendor recommends a product that considers the price response together, it will satisfy the customer's price preferences. Thus, the recommendation system can be evaluated to see if it provides price fairness as a useful evaluation scale for the user. According to Sergelen-Darhantoya, Park, and Lee (2019), a system for recommending products based on price fairness can be measured by the acknowledgment given by users when the product or service is priced appropriately and provides an added benefit.

Recommendation Quality

Personalization is a critical determinant of service quality (Mittal & Lassar, 1996). In this study, personalized recommendations contribute positively to the quality of services and users' experience. Users can assess the usefulness of the recommendation system by evaluating the quality provided by providers. Parasuraman et al. (1988) defined a service quality indicator as an "attitude or evaluation" of service quality. Lewis and Booms (1983) stated service quality as the degree to which the service conforms to the customer's expectations. In the context of the current study, in product recommendation systems, quality can be defined as the ability to evaluate the user's acceptance of the product and the consistency of its results with customer expectations.

Proposed Model

Figure 1. Proposed model

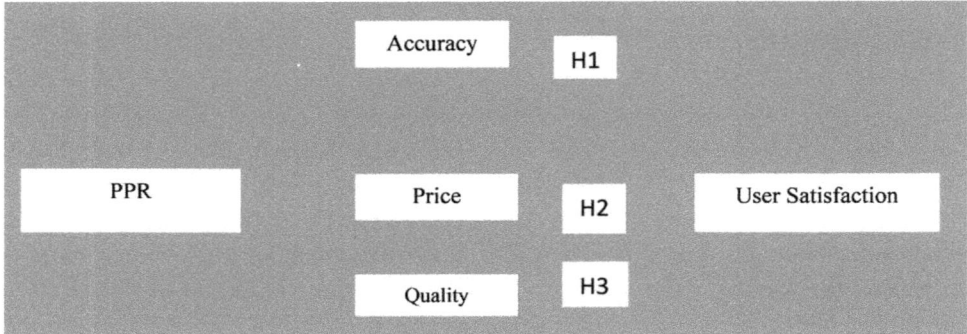

Based on previous studies and an extant literature review, this study proposed a model consisting of five propositions. These propositions are again subdivided into independent and dependent variables. PPR was the independent variable with three dimensions: accuracy, price, and quality. User satisfaction is considered the dependent variable in this model. As this study aimed to find out the significant association between product recommendations on customer satisfaction, hence by the help of the proposed hypothesis, the study tried to establish the relationship among the independent variables (recommendation accuracy, price of the recommended product, recommendation quality) and dependent variable (user satisfaction). PPR refers to personalized product recommendations.

Table 1. Measurement of research variables

Variables	Items	Scale
Accuracy	7	Herlocker (2004),
Price	5	Sergelen-Darhantoya, Park and Lee (2019)
Recommendation Quality	7	Benlian, et.al. (2012),
Satisfaction	2	Bhattacherjee (2001),

Research Hypotheses

This study examined the impact of product recommendations concerning customer satisfaction using respondents who have used a product recommendation for online shopping. Hence, an empirical analysis is being conducted to assess the effectiveness of online product recommendation systems on customer satisfaction. As per the literature review, the study postulated the following hypothesis,

H1: Customer satisfaction is positively affected by product recommendations accuracy.
H2: Fair-priced recommendation systems can lead to an increase in customer satisfaction.
H3: Customer satisfaction is positively affected by the quality of a product recommendation system.

Construct Measurements

As part of this study's research model, a product recommendation system was evaluated based on its characteristics and customer satisfaction. In this study, the Likert 7-point scale measures product recommendation systems' accuracy, quality, and price fairness aspects. The measurement of research variables is given in Table 1. This study adapted previous studies' measurements for all theoretical constructs. It also included a screening question that asked respondents whether they used OPRs during the past six months to purchase at least one recommended product. Several actions have been taken to ensure the reliability and validity of the survey instrument, including two expert panels, one e-retailer, and two online customers included in the pilot survey. During pilot testing, all constructs were internally consistent (alpha values were more significant than 0.8), and the survey questionnaire did not need to be modified further.

Research Method

To conduct the survey, primary and secondary data were gathered to study correlations between the variables. Primary data was collected with the help of an online survey. Survey participants had experience with online buying and were given access to a web-based survey. Before conducting the survey, content analysis was done, and a seven-point Likert scale was utilized to collect the data. A sample questionnaire survey was conducted to obviate any issues with this study. Exploratory factor analysis was performed to analyze the data further and illustrate the results.

Sample Design and Analysis Method

According to Amazon's findings, most people who purchase online have a high level of money, education, maturity, and a desire for a better quality of life since the 1980s and 1990s. Hence, in this study, survey participants are defined as anyone who has ever purchased online from a retail website to ensure the results' accuracy. Online questionnaires were sent to more than 265 surveys, and 214 were returned. Two hundred surveys were used after excluding the ones deemed unfit for inclusion. The majority of respondents in this study were born between 1980 and 1990. Table 2 shows that females comprised a smaller percentage of the population than males. The subject of this study was limited to the age group of 20 to 40-year-old female and male consumers who have used online shopping for the last three months. The survey was conducted online from June 15 to June 30 and involved 200 people. The online survey method was used to validate the validity of the research hypothesis and its suitability. The questionnaire was prepared for the survey by referring to the previous studies. A principal component analysis was conducted to refine the constructs, after which online product recommendation characteristics were confirmed. The study investigated reliability using Cronbach's alpha coefficient for each factor. Further, this study investigated the impact of product recommendations on customer satisfaction by their characteristics.

Empirical Analysis

General Characteristics of the Sample

According to demographic frequencies, 97 female respondents comprised 48.5 per cent of the entire sample, with 103 men making up the remaining 51.5 per cent. The age group from 15 to 20 represented 19.5 per cent of the total samples, 20 to 25 represented 20.5 per cent, 27 to 32 represented 17 per cent, and 33 to 38 represented 43 per cent. Online inexperienced buyers made up 12.5 per cent of the sample, those who were learning the ropes made up 17.5 per cent, those who were competently made up 36.5 per cent, and those who were experienced made up 33.5 per cent. The sample respondents' educational backgrounds included graduation, post-graduation, and other professional courses. Graduate students made up 39 per cent of the sample, postgraduate students made up 43.5 per cent, and students enrolled in professional courses made up 17.5 per cent of the total population. 23.5 per cent of the overall sample spent less than INR 2000 online shopping, while 34.5 per cent spent between INR 2000 and INR 4000 online shopping. Similarly, between INR 4000 and INR 5000, 24 per cent of the sample, more than INR 5000 or more, accounted for 18 per cent of the entire sample.

Table 2. Demographic profile of respondent

Category	Variable	Frequency	Percentage
Gender	Male	103	51.5
	Female	97	48.5
Age	15 -20	39	19.5
	21-26	41	20.5
	27-32 33-38	34 86	17 43
Monthly Expenditure	2000 and below	47	23.5
	2001 -4000	69	34.5
	4001 -5000	48	24
	5001 and above	36	18
Qualification Enrolled User Experience (online buying)	Graduate PG Professional course Novice Beginner Competent Expert	78 87 35 25 35 73 67	39 43.5 17.5 12.5 17.5 36.5 33.5

Reliability and Validity Analysis of Measurement

This survey used factor analysis to evaluate the validity of the measurement. The factors were extracted using the analysis, and the varimax rotation was used to put them into categories. According to Eigen Value 1, this study selected the number of factors to be extracted. Conceptually inconsistent items were not found during the analysis. The factors affecting customer satisfaction were quantified through the input of 21 measurement tools for the study concerning product recommendations. Further, the study identified the reliability of each variable through reliability analysis. Cronbach's alpha coefficient for the measured is 0.6 or more.

Table 3. Factor loadings

Factor	Item Loadings	Cronbach's α
Accuracy		.930
A1	.764	
A2	.705	
A3	.868	
A4	.946	
A5	.765	
A6	.804	
A7	.797	
Price		.914
P1	.733	
P2	.965	
P3	.925	
P4	.670	
P5	.926	
Satisfaction		.813
S1	.852	
S2	.516	
Recommendation Quality		.935
REQ1 **REQ2** **REQ3** **REQ4** **REQ5**	**.898** **.792** **.849** **.929** **.926**	

Table 4. Model summary

Model	R	R Square	Adjusted R Square	Std. Error of the Estimate
1	.682[a]	.465	.457	.6677

a. Predictors: (Constant), RQ, Price, Accuracy

Table 5. ANOVA

	Model	Sum of Squares	df	Mean Square	F	Sig.
1	Regression	76.063	3	25.354	56.865	.000[b]
	Residual	87.392	196	.446		
	Total	163.455	199			
a. Dependent Variable: Satisfaction						
b. Predictors: (Constant), RQ, Price, Accuracy						

Table 6. Regression analysis

Coefficients[a]						
Model		Unstandardized Coefficients		Standardized Coefficients	t	Sig.
		B	Std. Error	Beta		
1	(Constant)	.562	.453		1.240	.216
	Accuracy	.115	.038	.168	3.048	.003
	Price	.056	.043	.070	1.288	.000
	RQ	.759	.072	.589	10.517	.000
a. Dependent Variable: Satisfaction						

The findings of an analysis of the regression equation are as follows: R 2 =.465 accounts for 46.5 per cent of the total variation. The F value was 56.865, which at the .05 levels of significance was significant. According to the regression above, price, recommendation quality, and accuracy strongly correlate with customer satisfaction. This is seen from the calculated coefficient of the corresponding variables' positive signals. According to Table 6 for regression analysis and Table 4 of the summary model, the regression equation for user satisfaction with respect to the accuracy, price, and recommendation quality is as follows: (Satisfaction (S) = .562 + .115 accuracy + .056 price + .759 Recommendation Quality). Hence, the regression equation was significant, and hypotheses H1, H2, and H3 held true.

Practical Implications

Customer satisfaction is affected by all aspects of the product recommendation system, including price fairness, accuracy, and quality of recommendations. Among all three, it is observed that the beta value of recommendation quality is the highest among others. The above analysis clearly shows that product recommendation characteristics positively impact customer satisfaction, so a survey to measure quality needs to be conducted to fill the needs of consumers.

Measuring and assessing customers' satisfaction with recommended products is possible by evaluating their reactions after using them. Customers are encouraged to leave reviews on many online outlets' websites as part of the firm's customer relationship management strategy. Based on this need-rating information, customer satisfaction with the product can be estimated after its use. Through personal information and responses, etailers can better forecast customers' true feelings toward a particular product and suggest a more suitable offering for the potential customer.

This study found that personalized product recommendations significantly impact customer satisfaction. Therefore, to demonstrate OPRs' trustworthiness, online retailers need to maintain what measures are being taken to manage customers' expectations.

According to Wang and Head (2007), online repurchases are positively affected by satisfaction. It is more likely that satisfied customers will repurchase rather than dissatisfied customers (Garcia et al., 2012). Hence, ensuring customer satisfaction is imperative in online shopping decisions, increasing the likelihood of repeat purchases (Gupta & Kim, 2010). So, retailers should understand how to improve business performance and expand customer satisfaction levels in the online environment.

Managerial Implications

This research has significant implications for online retailers, researchers, and academics. This study has helped us understand the understudied topic of customer satisfaction with product recommendations for online shopping in India. Keeping a customer loyal to an online store depends on good customer service and support, so for an online store to succeed and remain loyal to its customers, customer service and support are vital factors (Ashraf, M. 2017). E-retailers have many new opportunities to improve the service they provide to customers thanks to advances in Web-based technologies. Consumers increasingly need help making buying decisions. Online retailers are increasingly providing their customers with highly personalized product recommendations on their e-commerce sites to aid them in buying decisions (Ashraf et al., 2020; Sheng et al., 2014).

Customers are encouraged to buy certain products when they receive online purchase recommendations (OPRs), which can increase spending and better customer retention. It can increase customer spending and retention by providing customers with a list of products they should purchase (OPRs). As long as the recommendations' quality, accuracy, and pricing are satisfactory to the customers, OPRs will likely continue to be used for future purchases. This indicates that OPR usage is acceptable when it meets customer expectations for better product evaluations. Customers will be most satisfied with OPR use if OPRs can fulfill their expectations for better product evaluation.

Customer satisfaction is among the most extensively researched topics in marketing (Lee & Whaley, 2019). It depends on how well the product or service fulfills or exceeds customers' needs or expectations (Prayag et al., 2019; Burns & Neisner, 2006). In addition, there is evidence linking satisfaction to loyalty and purchase intentions among customers (Ou & Verhoef, 2017; Lin et al.,2020; Park et al., 2019). A satisfied customer will likely become loyal; thus, marketers and retailers must develop strategies to retain customers by satisfying their needs. Also, e-retailers need to understand the relationship between recommendation features such as accuracy, quality, and price with customer satisfaction during the recommendation process.

Limitations and Scope of the Study

The survey was conducted with the help of 200 respondents, which needed to be more for statistical analysis. Therefore, future studies can be conducted by adding more respondents to draw new conclusions. Again, this study is limited to the age group of 15 to 40, which can further be expanded to the age group of more than 40 and will add to future research dimensions. Again, the study has considered product recommendation features for analysis which can add other features like customer reviews, ratings, product categories, and their impact on customer satisfaction. However, user convenience and perceived ease of use can be considered while considering product recommendations and user satisfaction.

Concluding Remarks

Industry 5.0 is the current revolution that emphasizes the use of advanced technologies combined with cognitive computing, with a return of human handwork as the focal point. It is as close to personalization as it can realistically get in a B2B world where AI provides customers with individualized shopping experiences and boosts customer experience. As a result, AI is helping to provide more efficient and improved customer retention. As a result, personalized recommendations are most likely to have high user satisfaction. This study's empirical results demonstrated a high degree of explanatory power, indicating that recommendation quality, accuracy, and price play a significant role in customer satisfaction. This would primarily be enhanced if OPRs met customer expectations regarding product evaluations. Following empirical findings and theoretical frameworks, this study proposed a conceptual model considering the constructs of personalized recommendation and consumer satisfaction and found that personalized recommendation positively influences customer satisfaction. Since high levels of customer satisfaction are related to customer loyalty, to perpetuate customer loyalty, boost brand reputation, and drive repeat business, marketers must maintain high customer satisfaction. This study shows that customer satisfaction measurement reflects the customer's attitude towards the accuracy, price, and quality of product recommendations, ultimately leading to purchase intention and customer loyalty. Hopefully, this study will provide new dimensions to customer satisfaction in the context of Industry 5.0 and personalized product recommendations.

REFERENCES

Agarwal, D. (2019). *Top 5 trends that will rule fashion retail in 2019*. Available at: www.indianretailer. com/article/whats-hot/trends/Top-5-trends-that-will-rule-fashion-retail-in-2019.a6270/

Agarwal, P., Vempati, S., & Borar, S. (2018). *Personalizing similar product recommendations in fashion e-commerce.* arXiv preprint arXiv:1806.11371.

Aggarwal, C. C. (2016). *Recommender Systems* (Vol. 1). Springer. doi:10.1007/978-3-319-29659-3

Ahluwalia, R., Unnava, H. R., & Burnkrant, R. E. (2001). The moderating role of commitment on the spillover effect of marketing communications. *JMR, Journal of Marketing Research*, *38*(4), 45870. doi:10.1509/jmkr.38.4.458.18903

Akundi, A., Euresti, D., Luna, S., Ankobiah, W., Lopes, A., & Edinbarough, I. (2022). State of Industry 5.0—Analysis and Identification of Current Research Trends. *Applied System Innovation*, *5*(1), 27. doi:10.3390/asi5010027

André, Q., Carmon, Z., Wertenbroch, K., Crum, A., Frank, D., Goldstein, W., Huber, J., van Boven, L., Weber, B., & Yang, H. (2018). Consumer choice and autonomy in the age of artificial intelligence and big data. *Customer Needs and Solutions*, *5*(1-2), 28–37. doi:10.100740547-017-0085-8

Ashraf, M., Ahmad, J., Hamyon, A. A., Sheikh, M. R., & Sharif, W. (2020). Effects of post-adoption beliefs on customers' online product recommendation continuous usage: An extended expectation-confirmation model. *Cogent Business & Management*, *7*(1), 1735693. doi:10.1080/23311975.2020.1735693

Ashraf, M., Jaafar, N. I., & Sulaiman, A. (2017). *The mediation effect of trusting beliefs on the relationship between expectation-confirmation and satisfaction with the usage of online product recommendation.* The South East Asian Journal of Management.

Balabanovic, M., & Shoham, Y. (1997). Fab: Content-based, collaborative recommendation. *Communications of the ACM, 40*(3), 66–72. doi:10.1145/245108.245124

Bathla, G., Aggarwal, H., & Rani, R. (2020). A graph-based model to improve social trust and influence for social recommendation. *The Journal of Supercomputing, 76*(6), 4057–4075. doi:10.100711227-017-2196-2

Bendapudi, N., & Berry, L. L. (1997). Customers' motivations for maintaining relationships with service providers. *Journal of Retailing, 73*(1), 15–37. doi:10.1016/S0022-4359(97)90013-0

Benlian, A., Titah, R., & Hess, T. (2012). Differential effects of provider recommendations and consumer reviews in E-commerce transactions: An experimental study. *Journal of Management Information Systems, 29*(1), 237–272. doi:10.2753/MIS0742-1222290107

Bhattacherjee, A. (2001). Understanding information systems continuance: An expectation-confirmation model. *Management Information Systems Quarterly, 25*(3), 351–370. doi:10.2307/3250921

Burns, D. J., & Neisner, L. (2006). Customer satisfaction in a retail setting: The contribution of emotion. *International Journal of Retail & Distribution Management, 34*(1), 49–66. doi:10.1108/09590550610642819

Changchun, S. W., Lee, C.-F., & Hsu, Y.-J. (2004). Online personalized sales promotion in electronic commerce. *Expert Systems with Applications, 27*(1), 35–52. doi:10.1016/j.eswa.2003.12.017

Chattopadhyay, S., Shankar, S., Gangadhar, R. B., & Kasinathan, K. (2018). Applications of artificial intelligence in assessment for learning in schools. In *Handbook of research on digital content, mobile learning, and technology integration models in teacher education* (pp. 185–206). IGI Global. doi:10.4018/978-1-5225-2953-8.ch010

Chen, T., Peng, L., Yin, X., Rong, J., Yang, J., & Cong, G. (2020, September). Analysis of user satisfaction with online education platforms in China during the COVID-19 pandemic. *Health Care, 8*(3), 200. PMID:32645911

Chen, Z., Ling, K. C., Ying, G. X., & Meng, T. C. (2012). Antecedents of online customer satisfaction in China. *International Business Management, 6*(2), 168–175. doi:10.3923/ibm.2012.168.175

Cheung, C. M. K., & Lee, M. K. O. (2012). What drives consumers to spread electronic word of mouth in online consumer-opinion platforms. *Decision Support Systems*, *53*(1), 21825. doi:10.1016/j.dss.2012.01.015

Cho, Y. H., Kim, J. K., & Kim, S. H. (2002). A personalized recommender system based on web usage mining and decision tree induction. *Expert Systems with Applications*, *23*(3), 329–342. doi:10.1016/S0957-4174(02)00052-0

Chung, K., & Shin, J. (2010). The antecedents and consequences of relationship quality in internet shopping. *Asia Pacific Journal of Marketing and Logistics*, *22*(4), 473–491. doi:10.1108/13555851011090510

Covington, P., Adams, J., & Sargin, E. (2016). Deep neural networks for youtube recommendations. *Proceedings of the 10th ACM conference on recommender systems*, 191–198. 10.1145/2959100.2959190

Cui, Z., Xu, X., Fei, X. U. E., Cai, X., Cao, Y., Zhang, W., & Chen, J. (2020). Personalized recommendation system based on collaborative filtering for IoT scenarios. *IEEE Transactions on Services Computing*, *13*(4), 685–695. doi:10.1109/TSC.2020.2964552

Cunningham, P., Bergmann, R., Schmitt, S., Traphoner, R., Breen, S., & Smyth, B. (2001). WebSell: Intelligent sales assistants for the World Wide Web. *Kunstliche Intelligenz*, *15*(1), 28–32.

Dash, G., Kiefer, K., & Paul, J. (2021). Marketing-to-Millennials: Marketing 4.0, customer satisfaction and purchase intention. *Journal of Business Research*, *122*, 608–620. doi:10.1016/j.jbusres.2020.10.016

Davenport, T., Guha, A., Grewal, D., & Bressgott, T. (2020). How artificial intelligence will change the future of marketing. *Journal of the Academy of Marketing Science*, *48*(1), 24–42. doi:10.100711747-019-00696-0

de Morais Watanabe, E. A., Torres, C. V., & Alfinito, S. (2019). *The impact of culture, evaluation of store image and satisfaction on purchase intention at supermarkets*. Revista de Gestão.

Dhingra, S., Gupta, S., & Bhatt, R. (2020). A study of relationship among service quality of E-commerce websites, customer satisfaction, and purchase intention. *International Journal of E-Business Research*, *16*(3), 42–59. doi:10.4018/IJEBR.2020070103

Di Nardo, M., & Yu, H. (2021). Special Issue "Industry 5.0: The Prelude to the Sixth Industrial Revolution". *Applied System Innovation*, *4*(3), 45. doi:10.3390/asi4030045

Dianat, I., Adeli, P., Jafarabadi, M. A., & Karimi, M. A. (2019). User-centred web design, usability and user satisfaction: The case of online banking websites in Iran. *Applied Ergonomics*, *81*, 102892. doi:10.1016/j.apergo.2019.102892 PMID:31422242

Doyle-Kent, M., & Kopacek, P. (2019). Industry 5.0: Is the Manufacturing Industry on the Cusp of a New Revolution? In *Proceedings of the International Symposium for Production Research* (pp. 432-441). Springer.

Eisenman, A., Naumov, M., Gardner, D., Smelyanskiy, M., Pupyrev, S., Hazelwood, K., Cidon, A., & Katti, S. (2018). *Bandana: Using nonvolatile memory for storing deep learning models.* arXiv preprint arXiv:1811.05922.

Eklof, J., Podkorytova, O., & Malova, A. (2020). Linking customer satisfaction with financial performance: An empirical study of Scandinavian banks. *Total Quality Management & Business Excellence*, *31*(15-16), 1684–1702. doi:10.1080/147833 63.2018.1504621

El-Adly, M. I. (2019). Modelling the relationship between hotel perceived value, customer satisfaction, and customer loyalty. *Journal of Retailing and Consumer Services*, *50*, 322–332. doi:10.1016/j.jretconser.2018.07.007

Elavarasan, R. M., Leoponraj, S., Vishnupriyan, J., Dheeraj, A., & Sundar, G. G. (2021). Multi-Criteria Decision Analysis for user satisfaction-induced demand-side load management for an institutional building. *Renewable Energy*, *170*, 1396–1426. doi:10.1016/j.renene.2021.01.134

Feldman, J. M., & Lynch, J. G. (1988). Self-generated validity and other effects of measurement on belief, attitude, intention and behavior. *The Journal of Applied Psychology*, *73*(3), 42135. doi:10.1037/0021-9010.73.3.421

Fornell, C. (1992). A national customer satisfaction barometer: The Swedish experience. *Journal of Marketing*, *56*(1), 6–21. doi:10.1177/002224299205600103

Fu, Y., Wan, J., Zhao, H., Jiang, W., & Pu, S. (2020). Preference-aware heterogeneous graph neural networks for recommendation. *2020 IEEE 32nd international conference on tools with artificial intelligence (ICTAI)*, 41–46. 10.1109/ICTAI50040.2020.00017

Ghasemaghaei, M., Hassanein, K., & Benbasat, I. (2019). Assessing the design choices for online recommendation agents for older adults: Older does not always mean simpler information technology. *Management Information Systems Quarterly*, *43*(1), 329–346. doi:10.25300/MISQ/2019/13947

Gomez-Uribe, C. A., & Hunt, N. (2015). The Netflix recommender system: Algorithms, business value, and innovation. *ACM Transactions on Management Information Systems*, *6*(4), 1–19. doi:10.1145/2843948

Greenwell, T. C., Fink, J. S., & Pastore, D. L. (2002). Assessing the influence of the physical sports facility on customer satisfaction within the context of the service experience. *Sport Management Review*, *5*(2), 129–148. doi:10.1016/S1441-3523(02)70064-8

Griffiths, J. R., Johnson, F., & Hartley, R. J. (2007). User satisfaction as a measure of system performance. *Journal of Librarianship and Information Science*, *39*(3), 142–152. doi:10.1177/0961000607080417

Grönroos, C. (1994). From scientific management to service management: A management perspective for the age of service competition. *International Journal of Service Industry Management*, *5*(1), 5–20. doi:10.1108/09564239410051885

Gupta, U., Wu, C. J., Wang, X., Naumov, M., Reagen, B., Brooks, D., ... Lee, H. H. S. (2020, February). The architectural implications of facebook's DNN-based personalized recommendation. In *2020 IEEE International Symposium on High Performance Computer Architecture (HPCA)* (pp. 488-501). IEEE. 10.1109/HPCA47549.2020.00047

Herlocker, J. L., Konstan, J. A., Terveen, L. G., & Riedl, J. T. (2004). Evaluating Collaborative Filtering Recommender Systems. *ACM Transactions on Information Systems*, *22*(1), 5–53. doi:10.1145/963770.963772

Holbrook, T. (1996). *Do campaigns matter?* (Vol. 1). Sage Publications. doi:10.4135/9781452243825

Hossain, M. S., Zhou, X., & Rahman, M. F. (2018). Examining the impact of QR codes on purchase intention and customer satisfaction on the basis of perceived flow. *International Journal of Engineering Business Management*, *10*. doi:10.1177/1847979018812323

Hsu, C. L., & Lin, J. C. C. (2015). What drives purchase intention for paid mobile apps?–An expectation confirmation model with perceived value. *Electronic Commerce Research and Applications*, *14*(1), 46–57. doi:10.1016/j.elerap.2014.11.003

Huang, Y., Wang, N., Zhang, H., & Wang, J. (2019). A novel product recommendation model consolidating price, trust and online reviews. *Kybernetes*, *48*(6), 1355–1372. doi:10.1108/K-03-2018-0143

Hui, B., Zhang, L., Zhou, X., Wen, X., & Nian, Y. (2022). Personalized recommendation system based on knowledge embedding and historical behavior. *Applied Intelligence*, *52*(1), 954–966. doi:10.100710489-021-02363-w

Iyengar, K. P., Pe, E. Z., Jalli, J., Shashidhara, M. K., Jain, V. K., Vaish, A., & Vaishya, R. (2022). Industry 5.0 technology capabilities in Trauma and Orthopaedics. *Journal of Orthopaedics*, *32*, 125–132. doi:10.1016/j.jor.2022.06.001 PMID:35707297

Javaid, M., & Haleem, A. (2020). Critical components of Industry 5.0 towards a successful adoption in the field of manufacturing. *Journal of Industrial Integration and Management*, *5*(3), 327–348. doi:10.1142/S2424862220500141

Jiang, Z., & Benbasat, I. (2004). Virtual product experience: Effects of visual and functional control of products on perceived diagnosticity in electronic shopping. *Journal of Management Information Systems*, *21*(3), 111–147. doi:10.1080/07421 222.2004.11045817

Karmarkar, U. R., Shiv, B., & Knutson, B. (2015). Cost conscious? The neural and behavioral impact of price primacy on decision-making. *JMR, Journal of Marketing Research*, *52*(4), 467–481. doi:10.1509/jmr.13.0488

Kim, H.-W., Xu, Y., & Gupta, S. (2012). Which is more important in internet shopping, perceived price or trust. *Electronic Commerce Research and Applications*, *11*(3), 241–252. doi:10.1016/j.elerap.2011.06.003

Kim, J.-K., Choi, I.-Y., & Li, Q. (2021). Customer Satisfaction of Recommender System: Examining Accuracy and Diversity in Several Type Recommendation Approach. *Sustainability (Basel)*, *13*(11), 6165. doi:10.3390u13116165

Kwon, Y., Park, J., & Son, J.-Y. (2021). Accurately or accidentally? Recommendation agent and search experience in over-the-top (OTT) services. *Internet Research*, *31*(2), 562–586. doi:10.1108/INTR-03-2020-0127

Kwon, Y., Son, S., & Jang, K. (2020). User satisfaction with battery electric vehicles in South Korea. *Transportation Research Part D, Transport and Environment*, *82*, 102306. doi:10.1016/j.trd.2020.102306

Lee, J., & Whaley, J. E. (2019). Determinants of dining satisfaction. *Journal of Hospitality Marketing & Management*, *28*(3), 351–378. doi:10.1080/19368623.2 019.1523031

Lee, M.-C. (2010). Explaining and predicting users' continuance intention toward e-learning: An extension of the expectation–confirmation model. *Computers & Education*, *54*(2), 506–516. doi:10.1016/j.compedu.2009.09.002

Lee, S. J., & Lee, H. C. (2007). A Study on Prediction Performance of Correspondence Average Algorithm in Cooperative Filtering Recommendation. *InformationSystem Review, 9*(1), 85–103.

Lewis, R. C., & Booms, B. H. (1983). The Marketing Aspects of Service Quality. In L. Berry, L. Shostack, & G. Upah (Eds.), *Emerging Perspectives on Services Marketing* (pp. 99–107). American Marketing Association.

Liang, T., Lai, H., & Ku, Y. (2007). Personalized Content Recommendation and User Satisfaction: Theoretical Synthesis and Empirical Findings. *Journal of Management Information Systems, 23*(3), 45–70. doi:10.2753/MIS0742-1222230303

Lin, H., Zhang, M., & Gursoy, D. (2020). Impact of nonverbal customer-to-customer interactions on customer satisfaction and loyalty intentions. *International Journal of Contemporary Hospitality Management, 32*(5), 1967–1985. doi:10.1108/IJCHM-08-2019-0694

Liu, L., Cheung, C. M. K., & Lee, M. K. O. (2016). An empirical investigation of information sharing behavior on social commerce sites. *International Journal of Information Management, 36*(5), 686–699. doi:10.1016/j.ijinfomgt.2016.03.013

Liu, Y., & Jang, S. S. (2009). Perceptions of Chinese restaurants in the US: What affects customer satisfaction and behavioral intentions? *International Journal of Hospitality Management, 28*(3), 338–348. doi:10.1016/j.ijhm.2008.10.008

Maddikunta, P. K. R., Pham, Q. V., Prabadevi, B., Deepa, N., Dev, K., Gadekallu, T. R., ... Liyanage, M. (2022). Industry 5.0: A survey on enabling technologies and potential applications. *Journal of Industrial Information Integration, 26*, 100257. doi:10.1016/j.jii.2021.100257

Martynov, V. V., Shavaleeva, D. N., & Zaytseva, A. A. (2019). Information Technology as the Basis for Transformation into a Digital Society and Industry 5.0. In *2019 International Conference Quality Management, Transport and Information Security, Information Technologies (IT&QM&IS)* (pp. 539-543). IEEE.

Maslowska, E., Smit, E. G., & van den Putte, B. (2016). It is all in the name: A study of consumers' responses to personalized communication. *Journal of Interactive Advertising, 16*(1), 74–85. doi:10.1080/15252019.2016.1161568

Mittal, B., & Lassar, W. M. (1996). The role of personalization in service encounters. *Journal of Retailing, 72*(1), 95–109. doi:10.1016/S0022-4359(96)90007-X

Mudambi, S. M., & Schuff, D. (2010). What makes a helpful online review? A study of customer reviews on Amazon.com. *Management Information Systems Quarterly*, *34*(1), 185200. doi:10.2307/20721420

Nassar, N., Jafar, A., & Rahhal, Y. (2020). A novel deep multi-criteria collaborative filtering model for recommendation system. *Knowledge-Based Systems*, *187*, 104811. doi:10.1016/j.knosys.2019.06.019

Naumov, M. (2019). *On the dimensionality of embeddings for sparse features and data*. arXiv preprint arXiv:1901.02103.

Oliver, R. L. (1997). A Conceptual Modes of Service Quality and Service Satisfaction: Compatible Goals, Different Concepts. *Advance in Services Marketing and Management, 2*, 65–85.

Ostergaard, E. H. (2018). *Welcome to industry 5.0*. Retrieved from https://isajobs. isa.org/intech/20180403/

Ou, Y. C., & Verhoef, P. C. (2017). The impact of positive and negative emotions on loyalty intentions and their interactions with customer equity drivers. *Journal of Business Research*, *80*, 106–115. doi:10.1016/j.jbusres.2017.07.011

Ozdemir, V., & Hekim, N. (2018). Birth of Industry 5.0: Making sense of big data with artificial intelligence, "the Internet of things" and next-generation technology policy. *OMICS: A Journal of Integrative Biology*, *22*(1), 65–76. doi:10.1089/omi.2017.0194 PMID:29293405

Palma, D., Ortúzar, J. D. D., Rizzi, L. I., Guevara, C. A., Casaubon, G., & Ma, H. (2016). Modelling choice when price is a cue for quality a case study with Chinese wine consumers. *Journal of Choice Modelling*, *19*, 24–39. doi:10.1016/j.jocm.2016.06.002

Parasuraman, A., Berry, L. L., & Zeithaml, V. A. (1993). More on improving service quality measurement. *Journal of Retailing*, *69*(1), 140–147. doi:10.1016/S0022-4359(05)80007-7

Parasuraman, A., Zeithaml, V.A. & Berry, L.L. (1988). SERVQUAL: A multiple-item scale for measuring consumer perceptions of service quality. *Journal of Retailing*, *64*, 12-40.

Park, J. Y., Back, R. M., Bufquin, D., & Shapoval, V. (2019). Servicescape, positive affect, satisfaction and behavioral intentions: The moderating role of familiarity. *International Journal of Hospitality Management*, *78*, 102–111. doi:10.1016/j.ijhm.2018.11.003

Pfeiffer, J., & Scholz, M. (2013). A Low-Effort Recommendation System with High Accuracy. *Business & Information Systems Engineering*, *5*(6), 397–408. doi:10.100712599-013-0295-z

Prayag, G., Hassibi, S., & Nunkoo, R. (2019). A systematic review of consumer satisfaction studies in hospitality journals: Conceptual development, research approaches and future prospects. *Journal of Hospitality Marketing & Management*, *28*(1), 51–80. doi:10.1080/19368623.2018.1504367

Salameh, A., Hatamleh, A., Azim, M., & Kanaan, A. (2020). Customer oriented determinants of e-CRM success factors. *Uncertain Supply Chain Management*, *8*(4), 713–720. doi:10.5267/j.uscm.2020.8.001

Santa, R., MacDonald, J. B., & Ferrer, M. (2019). The role of trust in e-Government effectiveness, operational effectiveness and user satisfaction: Lessons from Saudi Arabia in e-G2B. *Government Information Quarterly*, *36*(1), 39–50. doi:10.1016/j.giq.2018.10.007

Sergelen, D., Park, Y. S., & Lee, D. S. (2019). The Effect of Trust in Online Shopping Mall and Trust in Recommendation System on Cross-Purchase Intention. *The Articles of the Korean Industrial Engineering Society's Spring Joint Conference*, *4*, 4821-4834.

Shahbazi, Z., & Byun, Y. C. (2019). Product recommendation based on content-based filtering using XGBoost classifier. *Int. J. Adv. Sci. Technol*, *29*, 6979–6988.

Shahbazi, Z., Hazra, D., Park, S., & Byun, Y. C. (2020). Toward improving the prediction accuracy of product recommendation system using extreme gradient boosting and encoding approaches. *Symmetry*, *12*(9), 1566. doi:10.3390ym12091566

Sheng, X., Li, J., & Zolfagharian, M. A. (2014). Consumer initial acceptance and continued use of recommendation agents: Literature review and proposed conceptual framework. *International Journal of Electronic Marketing and Retailing*, *6*(2), 112–127. doi:10.1504/IJEMR.2014.066467

Tandon, U., Kiran, R., & Sah, A. (2017). Analyzing customer satisfaction: Users perspective towards online shopping. *Nankai Business Review International*, *8*(3), 266–288. doi:10.1108/NBRI-04-2016-0012

Venkatesh, V., Morris, M. G., Davis, G. B., & Davis, F. D. (2003). User acceptance of information technology: Toward a unified view. *Management Information Systems Quarterly*, *27*(3), 425–478. doi:10.2307/30036540

Walia, N., Srite, M., & Huddleston, W. (2016). Eyeing the web interface: The influence of price, product, and personal involvement. *Electronic Commerce Research*, *16*(3), 297–333. doi:10.100710660-015-9200-9

Wang, W., & Wang, M. (2019). Effects of sponsorship disclosure on perceived integrity of biased recommendation agents: Psychological contract violation and knowledge-based trust perspectives. *Information Systems Research*, *30*(2), 507–522. doi:10.1287/isre.2018.0811

Wang, Y., Han, L., Qian, Q., Xia, J., & Li, J. (2022). Personalized recommendation via multi-dimensional meta-paths temporal graph probabilistic spreading. *Information Processing & Management*, *59*(1), 102787. doi:10.1016/j.ipm.2021.102787

Wang, Y., Tang, T., & Tang, J. (2001). An instrument for measuring customer satisfaction toward web sites that market digital products and services. *Journal of Electronic Commerce Research*, *2*(3), 89–102.

Wen, J., Zhu, X. R., Wang, C. D., & Tian, Z. (2022). A framework for personalized recommendation with conditional generative adversarial networks. *Knowledge and Information Systems*, *64*(10), 2637–2660. Advance online publication. doi:10.100710115-022-01719-z

Wu, H. (2021, February). Application of Collaborative Filtering Personalized Recommendation Algorithms to Website Navigation. *Journal of Physics: Conference Series*, *1813*(1), 012048. doi:10.1088/1742-6596/1813/1/012048

Xiao, B., & Benbasat, I. (2014). Research on the Use, Characteristics, and Impact of e-Commerce Product Recommendation Agents: A Review and Update for 2007-2012. In Handbook of Strategic eBusiness Management (pp. 403-431). Springer Berlin Heidelberg.

Xiao, B., & Benbasat, I. (2015). Designing Warning Messages for Detecting Biased Online Product Recommendations: An Empirical Investigation. *Information Systems Research*, *26*(4), 793–811. doi:10.1287/isre.2015.0592

Xiao, B., & Benbasat, I. (2018). An empirical examination of the influence of biased personalized product recommendations on consumers' decision making outcomes. *Decision Support Systems*, *110*, 46–57. doi:10.1016/j.dss.2018.03.005

Xu, J., Benbasat, I., & Cenfetelli, R. T. (2014). The Nature and Consequences of Trade-off Transparency in the Context of Recommendation Agents1. *Management Information Systems Quarterly*, *38*(2), 379–406. doi:10.25300/MISQ/2014/38.2.03

Yi, C., Jiang, Z., & Benbasat, I. (2017). Designing for diagnosticity and serendipity: An investigation of social product-search mechanisms. *Information Systems Research, 28*(2), 413-429.

Yiu, C.S., Grant, K., & Edgar, D. (2007). Factors affecting the adoption of internet banking in Hong Kong – implications for the banking sector. *International Journal of Information Management, 27*(5), 336–351.

Zhang, H., Wang, Z., Chen, S., & Guo, C. (2019). Product recommendation in online social networking communities: An empirical study of antecedents and a mediator. *Information & Management, 56*(2), 185–195. doi:10.1016/j.im.2018.05.001

Section 4
Industry 5.0: Transforming Learning to Education 5.0

Chapter 7
Education 5.0 Serving Future Skills for Industry 5.0 Era

Mahmoud Numan Bakkar
Institute of Applied Technology, Abu Dhabi Vocational Education and Training Institute, UAE

Arshia Kaul
Carpediem EdPsych Consultancy LLP, India

ABSTRACT

This chapter addresses the relationship between Industry 5.0 and Education 5.0, outlines the future skills required for Industry 5.0, and illustrates how these two sectors interact with the Sustainable Development Goals (SDG) for modernizing Society 5.0. It places a strong emphasis on high-quality education and uses tailored learning to help employees and companies optimize for Industry 5.0. Additionally, it discusses the OECD Learning Compass 2030 and Future of Education and Skills 2030 reports, as well as the International Organization for Standardization (ISO) 21001:2018-ISO standard utilized for future educational management. The chapter emphasizes the need to improve the educational system while presenting various options for adopting personalized education.

INTRODUCTION

Various disciplines are being explored by Industry 5.0 (I 5.0), including artificial intelligence-related areas and educational and social studies. Utilizing new technology in planning and designing highly customized and personalized education programs is the main goal of I 5.0 in education. This chapter will look at opportunities for Industry 5.0 tools in education. The chapter will examine topics from different educational

DOI: 10.4018/978-1-7998-8805-5.ch007

and academic needs to reach that goal. It will start with humanizing the education needs, explaining the meaning of Society 4.0, presenting the educational skills needed for the Industry 5.0 era, and showcasing the standards by the International Organization for Standardization adapted to Education 5.0 explaining the strategic alignment between SDG (Sustainable Development Goals) and Education 5.0. The chapter will also, discuss the importance of personalized education design as a major concept in Industry 5.0.

In all aspects of our lives, including our educational journey, Industry 4.0 leads to more automation, sophisticated Robots, IoT solutions, and Artificial Intelligence (AI) solutions as we move forward. The question that arises is whether we are losing the human touch. This brings forward the I 5.0 concept. Societies that developed from Society 1.0 evolved on hunting to Society 2.0, dependent on agriculture and advanced to Society 3.0, dependent on the industry that grew from Industry 1.0, based on mechanization using steam, water, and fossil fuels (Cabinet Office, 2023). Industry 2.0 was about the mass production approach of electricity. Industry 3.0 was about automation and information technology, which evolved into Industry 4.0. Cyber-physical systems and technologies such as AI, IoT, machine learning and big data analytics emerged as the Industry 5.0 revolution focusing on human, personal design for massive customized productions and co-robot collaboration (Saxena et al., 2020). In the Industry 5.0 era, the focus is shifting from digitization to personalization. In parallel, the educational systems evolved from Education 1.0, which used to be normal classroom teaching using basic methodologies, moving to Education 2.0, which involved research contributions as an added component in teaching. After that, the educational systems emphasized community engagement by adding community services in its systems design. Success in Industrial 4.0 technology made innovation a key skill in Education 4.0. Then finally, Education 5.0 educational systems concern massive personalization leading to massive industrialization (Saxena et al., 2020). Everything is smart in this society, so it needs a smart educational approach. Education 5.0 in the Industry 5.0 era aims to have personalized human aspects considered in educational program design, development, delivery, and effectiveness measurement.

In 2016 according to Lynch (2018), the book Humanizing the Education Machine concerned how education machines can be humanized. This book was about instructional models that focussed on learning intrinsic humanized values. These models promoted design thinking by improving classroom teaching and having models centered on the student, giving an example of a school rated above the state average. According to Lynch (2018), schools adopted science, technology, engineering, and mathematics (STEM) classrooms, which improved communication, collaboration, creative thinking, and critical thinking. In addition, the classrooms were designed with six tables with flat screen monitors used by small groups of students.

However, the cyber-physical relationship was built through Industry 4.0 paradigms that evolved into machine consciousness of the environment. These machines used sophisticated sensors to collect data processed with machine learning algorithms to generate knowledge that was evaluated based on a set of quality assurance attributes, such as: accuracy; confusion matrix; sensitivity; specificity; precision; and recall. These attributes are examples used in artificial intelligence (AI) to measure the performance level of the AI solution. After that, there was useful customization and personalization with a real human touch of added value. This enhanced feature gave rise to a more human-centered design product, which evolved with Industry 4.0 technology. Lantada (2022) mentioned that Society 5.0, announced in Japan, was human-centered. A society that balances economic advancement with social problem resolution by integrating cyber-space and physical space, as shown in Figure 1. Finally, the chapter will present examples of future skills demonstrations from different regions of the world.

Figure 1. Society 5.0
Source: Adapted from Cabinet Office (2023)

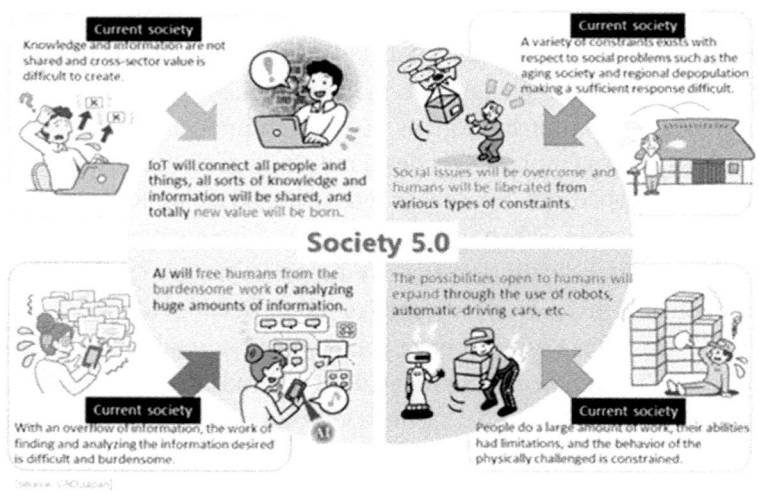

EDUCATIONAL SKILLS IN INDUSTRY 5.0

What Is Skill?

There is a comprehensive view of what a skill is. According to the Oxford Learners Dictionaries (2023), it is the ability to do something well. Yet, the IGI Global Dictionary (2023) defines skills differently, including: abilities resulting from training, experience, practice, or talent. This IGI dictionary entry also describes the ability to apply knowledge and demonstrate the know-how to accomplish tasks and solve problems. Skills are also defined as a synonym of competency, and its lifelong learning, special type of capability, and the ability to reach a specific result within a set time or energy according to the Oxford Learners Dictionaries (2023). Skills can also be grouped into domain-general skills such as teamwork, time management, leadership, and self-motivation (Oxford Learners Dictionaries, 2023). Domain-specific skills are needed for a specific job (Oxford Learners Dictionaries, 2023). Skills can also be classified by job function. If it involves ideas, it requires cognitive skills; if it involves things, it needs technical skills. If it affects people, it requires interpersonal skills. Skills need the knowledge to be applied. For example, a student could use it for science, technology, engineering, and mathematics (STEM), graduation or a capstone project. Skills refer to the functional ability to perform an action or do something (Oxford Learners Dictionaries, 2023).

Yet again, in another view, Güğerçin (2021) classified the top ten skills into four categories:

1. Problem-solving;
2. Working with people;
3. Technology use and development; and
4. Self-management.

It was proposed by Ehlers and Kellermann (2019), that education in the future will be active learning that caters to the student's choice and autonomy. It will have a structural change, supporting credit transfer for learning modules, micro-credentials, and qualifications. Also, academic learning will have a pedagogical design, such as a new assessment design and higher education. This will enable students to personalize their curriculum, gain lifelong learning, with graduate attributes focusing on future skills, and provide multi-institutional experience in their study (Ehlers & Kellermann, 2019).

The needed skills in Industry 5.0, represented in a project named Future Skills, vary from country to country. Some international organizations put their input on predicting future skills, such as the LinkedIn network and Coursera online platform;

both prepared a report summarizing the forecasted needed skills. In addition, countries like the UK, USA, UAE, and European Union emphasize preparing students for future skills. Table 2 summarizes the predicted future skills per a selected example of countries and organizations.

The United Arab Emirates (UAE) and the USA demonstrate the potential planning and future prediction of industry 5.0 competencies in education 5.0. The UAE's first future skills prediction is shining, an initiative organized by the Abu Dhabi Center for Technical and Vocational Education and Training (ACTVET), the national arm of vocational education and training in the UAE. The initiative, called Emirates Skills, conducts an annual competition including high-level applied technology students from the Abu Dhabi Polytechnic (AD Poly) Institute; Abu Dhabi Vocational and Educational Institute (ADVETI); and Applied Technology High School (ATHS) students (Emirates Skills, 2023). The initiative included the following technical skills categorized in its 2023 version as shown on its public website https://www.emiratesskills.ae:

1.   Construction and Building Technology;
2.   Manufacturing and Engineering Technology;
3.   Transportation and Logistics;
4.   Creative Arts and Fashion;
5.   Information and Communication Technology; and
6.   Social and Personal Services.

Emirates Skills (2023) promotes the following competitions as a tool to foster skill levels UAE youth:

- Skills Challenge;
- The National Competition;
- World Skills GCC Competition; and
- World Skills Competition.

Another example is Skills USA (see www.skillsusa.org). In promoting their skills, they focus on creating the Skills USA framework, illustrated in Table 1, categorizing the skills into 1- personal skills, 2- workplace skills, and 3- technical skills grounded in academics (Skills USA, 2023).

The adoption of these skills begins in the classroom, chapters, and clubs in school and the workplace. Another outlook on future skills comes from international organizations such as the British Council and Coursera's online education platforms. For example, it's expected after 15 years that, robots will service 250,000 UK public sector jobs (Council, 2018). A study conducted by the British Council in the UAE empathizes with the need to set up a platform that integrates the private sector, which is the employers and the workplaces, with the educators to help update the curriculum and the dynamic needed skills (Council, 2018). Coursera's online platform classified skills into three types (Business, Technology, and Data Science), and the company provided a sample of online courses that help close skill gaps (Sands et al., 2020). Another outlook on future skills from Singapore classified the skills into three main domains: Green Economy, Digital Economy, and Care Economy Skills (Skills Future Singapore, 2022).

Table 2 summarizes the skills needed in the future, recommended by the organizations mentioned above and Kingston University-UK.

Table 1. Skills USA framework

Personal Skills	Workplace Skills	Technical Skills Grounded in Academics
Integrity	Communication	Computer
Work Ethics	Decision-Making	Technology and Literacy
Professionalism	Teamwork	Job-Specific Skills
Responsibility	Multicultural Sensitivity	Safety and Health
Adaptability/Flexibility	Awareness	Service Orientation
Self-Motivation	Planning, Organizing, Management, and Leadership	Professional Development

Source: Adapted from Skills USA (2023)

Embedded Industry 5.0 Concepts in Education 5.0

Industry 5.0 main concepts: personalized design, co-robots, and human-centered, will be utilized in Education 5.0 in different approaches and ideas. In personalized education, learning experiences are tailored to student's unique learning preferences, strengths, and interests. In Industry 5.0, skills require intensive personalization in education. Here are some personalized educational ideas:

- **Adaptive Learning Technology:** Adaptive learning technology uses data and algorithms to adapt the pace and content of learning to the individual needs of each student. It can help students learn at their own pace and provide real-time feedback to teachers about student progress (Muñoz et al., 2022);

Table 2. Future skills

British Council - UAE	Coursera - Global			Kingston University UK	Singapore		
	Business	Technology	Data Science		Green Economy Skills domains	Digital Economy Skills Domains	Care Economy Skills Domains
Complex Problem-Solving							
Critical Thinking	Accounting Sample skills: Auditing, Financial Accounting	Computer Networking Sample skills: Cloud Computing, Internet of Things	Data Management Sample skills: Cloud APIs, Hadoop	Problem-solving/ process skills	Environmental and Sustainability Management	AI, Data, and Analytics	Person-Centered Care
Creativity	Communication Sample skills: People Skills, Writing	Databases Sample skills: Relational Database, Key-Value Database	Data Visualization Sample skills: Tableau, Plotting Data	Critical thinking	Green Infrastructure and Mobility	E-commerce and Digital Marketing	Collaboration with Stakeholders
People Management	Finance Sample skills: Financial Ratios, Blockchain	Operating Systems Sample skills: Mobile App Development, C Programming Language	Machine Learning Sample skills: Multi-Task Learning, Deep Learning	Communication skills	Energy, Resource Circularity and Decarbonization	Cybersecurity and Risk	Teaching and Learning
Coordinating with Others	Management Sample skills: People Management, Business Analytics	Security Engineering Sample skills: Cybersecurity, Cryptography	Math Sample skills: Calculus, Linear Algebra	Digital skills	Sustainable Finance	Cloud, Systems, and Infrastructure	Health and Wellness
Emotional Intelligence	Marketing Sample skills: Digital Marketing, Product Placement	Software Engineering Sample skills: Software Architecture and Development	Statistical Programming Sample skills: R, Python	Analytical skills		Software Development	
Judgment and decision-making	Sales, Sample skills: Cross-Selling, Lead Generation	Computer Programming Sample skills: JavaScript, Java	Statistics Sample Skills: Regression and AB Testing	Adaptability		Technology Application and Management	

continues on following page

Table 2. Continued

British Council - UAE	Coursera - Global		Kingston University UK	Singapore
Service orientation	Entrepreneurship Sample Skills: Adaptability and Innovation	Theoretical Computer Science Sample Skills: Algorithms and Cryptography	Data Analysis Sample skills: Exploratory Data Analysis, Spatial Data Analysis	Resilience	
Negotiation	Strategy and Operations Sample skills: Operations Management, Strategy	Cloud Computing Sample skills: Software as a Service, Kubernetes		Creativity	
Cognitive flexibility	Human Resources Sample Skills: Benefits and Employee Relations	Web Development Sample skills: Angular, HTML, and CSS		Ability to build relationships	
		Mobile Development Sample skills: Android Development, iOS Development		Initiative	

Source: Adapted from Council (2018), Sands et al. (2020), Future Skills (2022), and SkillsFuture Singapore (2022)

- **Student-Led Conferences:** Instead of traditional parent-teacher conferences, student-led conferences allow students to lead the conversation about their learning progress. As a result of this approach, students are empowered to take ownership of their learning and could reflect on their strengths, challenges, and areas for improvement (Conderman et al., 2000). A flexible schedule allows students to learn at their own pace and choose when and where they want to learn. Students with outside commitments or who learn best at specific times can benefit from this option;
- **Individualized Instruction:** Individualized instruction involves tailoring lessons and assignments to meet the unique needs of each student. This approach can help students who are struggling with a particular concept or who need more challenging material to stay engaged and motivated (Butler, 2002).
- **Multi-Age Classrooms:** These classrooms combine students ranging in age and grade level. This approach allows students to work with and learn from peers who have different skills, interests, and backgrounds (Hallion, 1994). A personal learning network enables students to connect with peers, mentors, and experts with similar interests. It can provide students with resources and support that can assist them in pursuing their goals and interests (Richardson & Mancabelli, 2011).
- **Project-Based Learning (PBL):** Project-based learning allows students to work on a project or inquiry-based activity that aligns with their interests and strengths. This approach helps students develop critical thinking, problem-solving, and communication skills while learning about a relevant topic (Lantada, 2022).

The above ideas play a crucial role in transforming knowledge into skills and integrating knowledge and skills into real-life applied applications. Hence, the need for human touch and further personalization and customization set with innovation and creativity. It's expected that by 2025 the industry 5.0 implementation plans will start.

And so transferring knowledge to applied skills needs to be framed in an educational training program. This activity fills the expected skills gap between current and anticipated skills. For example, Bakkar (2016) aligned knowledge measurements with task achievement through the progress of three consecutive knowledge cognition levels, declarative knowledge (verbal information skills, intellectual skills), procedural (intellectual skills, cognitive strategy), and meta-cognitive knowledge.

The International Bureau of Education and UNESCO material for training tools for curriculum development (IBE-UNESCO, 2017) clearly defined personalized learning. They have defined personalized learning as the fact that education needs to be personal to be effective. Any instruction to be relevant in the future must be application-oriented, and the learners must take full ownership of their learning. Facilitators cannot make education happen for someone, strategies only be used to enthuse learners to learn, and they can only make sense when they take ownership of it (IBE-UNESCO, 2017).

Personalized learning is essentially learner-centered education. It is very different from the earlier teacher-talk model, where the teachers used to be the main ones talking, and they would give long lectures with minimum interaction. It is also different from a textbook approach, where the teachers are there only to replicate the textbook exercises. Instead, personalized learning ensures that students take ownership of their work and must become lifelong learners.

In summary, the concept of personalized learning requires: (1) key ideas of relevance, inquiry, and differentiation; (2) there are different types of learners, different strategies for storing and making sense of the information at hand, and different ways in which the brain works for different learners; (3) the new technologies have acted as enhancers to personalized learning; and (4) UNESCO aims to see that the access to quality education means access to personalized learning (IBE-UNESCO, 2017).

For the system to align for personalized learning, the following are required:

1. Everyone should have the same understanding of personalized learning. All stakeholders, whether they be teachers, administrators, parents, leaders, or students, must have the same understanding of the concept. They should also believe that personalized learning is the best way to go forward;
2. At every stage, one must keep understanding the needs of the students;
3. The environment in which the learners' voice is heard and prepared to act upon what the learners, who are sometimes children, want;
4. Re-visiting goals and strategies continuously is an option for the system;
5. It is a system in which mistakes can be made, and it is taken to be a step toward learning. Students and learners should be comfortable acknowledging that they did not understand a concept taught and given another chance to understand;
6. The new technologies are ways in which learners can enhance their learning, they do not fully rely on them, but it is one way of doing so;
7. There is always access to students and teachers for learning materials to explore more about the domain knowledge they want;
8. Complete coherent curriculum, assessment, teaching, and learning with personalized learning; and
9. A system that tolerates outdoor education, field trips, and community and service activities.

Sustainable Development Goals and Education 5.0

The previously mentioned ideas for personalized education will assist crucially in transforming knowledge and skills into applications and student-centered education. Having ideas aligned with the sustainable development goals (SDG) strategically aligned with Education 5.0 is a crucial requirement in Education 5.0, as shown in Figure 2. In addition to using motivation and awarding challenges such as national and international competitions and awards as a goal, students working to achieve them could act as a performance indicator for the success of the personalized curriculum design in Education 5.0. Education 5.0 classroom design based on the personalized design as per the students' individual needs could be achieved using different educational tools. For example: problem based learning (PBL); flipped classrooms; using games in education; online applications and tools; community engagement; industrial association collaboration; and engagement such as the Institute of Electrical and Electronics Engineers (IEEE); Accreditation Board for Engineering and Technology; Association to Advance Collegiate Schools of Business; and the Australian Computer Society (ACS). Education 5.0 aims to utilize Industrial 4.0 technologies to provide high-quality education aligned with SDG Goals for servicing society 5.0 (Lantada, 2022).

Figure 2. Sustainable development goals
Source: Adapted from United Nations (2015)

Sustainable Development Goal 4 is related to quality education. It helps to define that the goal is equitable quality education and to promote lifelong learning opportunities for everyone. Education is believed to liberate intellect, make people imaginative, and helps them gain self-respect. Therefore, it is one of the key elements of prosperity and gives several opportunities, making it possible for everyone to be progressive and be part of a healthy society (The Global Goals, 2015).

Under this main goal of quality education, some goals have been stated to achieve the main goal. The aim is that: by 2030 all girls and boys must have free primary and secondary education; all students must have access to good quality early childhood development; there should be equal access to affordable technical, vocational, and higher education for all men and women; there should be an increase in the number of people with relevant skills by 2030, these skills include vocational skills for employment and other projects; all discrimination must be eliminated, and all youth must have literacy and numeracy by 2030; build citizens for the future with the knowledge and skills for sustainability in the future; build and upgrade inclusive and safe schools; increase the number of qualified teachers in developing countries. These are the goals set under the UN sustainable development goals (Global Goals, 2015).

The question remains, how to assess whether educational institutions have achieved quality education? One way is to establish a certain standard for measuring to judge whether the required criteria are met. One of the recent quality standards for the development of educational organizations is given by the International Organization for Standardization (ISO). This standard is ISO 21001:2018. Therefore, it is necessary to understand this standard in detail and what is expected of educational institutions in their functioning. In the following section, we explain ISO 21001:2018.

ISO 21001:2018

It is one of the latest standards published by the International Organization for Standardization -ISO for the Management Systems for Educational Organizations (EOMS) (PECB University, 2018). It is a standard that focuses on the services and products and the related improvement and or enhancement of educational services and products.

It is supposed to be a management tool to help educational institutions set up their policies and procedures so that all the stakeholders' needs and objectives are met, whether they are students, teachers, or other staff. This standard is in line with other ISO standards and is now well-known. ISO 21001 is a standard that would want to provide management tools for organizations to provide educational products and services to meet the needs of the beneficiaries. If an organization wants to comply with these standards, they would: look closely at internal auditing and learner satisfaction evaluations; control the external processes; products; or related services; review programs; and always try to address gaps in providing continuous improvement (PECB University, 2018).

Some of the principles of the EOMS are: (1) focus must always be on learners and beneficiaries; (2) the leadership should be visionary which would consider learner needs and set the organizational objectives accordingly; (3) engage people in a way that all are competent in delivering good value education; (4) process approach, which would consider the continuous improvement in processes; (5) improvement of overall organization at every stage; (6) the changes in the systems are taken based on data and information as proof; (7) the educational organizations should also focus on developing relations with all stakeholders; (8) the organizations must be socially responsible as well; (9) the organizations must also be able to provide for special needs of learners and provide an accessible environment; (10) the conduct related to educational standards must be ethical; and (11) there should be the protection of data in educational organizations (PECB University, 2018).

OECD Future of Education and Skills 2030

Finally, OECD (2018) launched the projects OECD Future of Education and Skills 2030 and OECD Learning Compass 2030 to assist future students in finding their educational pathway. As a framework for learning, the compass helps with recognizing formal and informal learning in addition to structured curriculums guided by instructional strategies (OECD, 2018). OECD distinguished between traditional education and the new normal of education using different features such as the student in the new normal education is engaged and shared the responsibility with other stakeholders such as parents, communities, and employers; it emphasizes the student's well-being not only on the student's performance and the educational outcomes, to evaluate the quality and the effectiveness of the learning delivery, also the curriculum is not linear designed, and it's enabling the students to have their learning path, in addition to the continuous improvement using the sufficient feedback and monitoring to achieve that, the needs for different assessments is crucial, and the student's role is to be active among all stakeholders (OECD, 2018)

The Organization for Economic Co-operation and Development (OECD) Learning Framework 2030 offers a framework for the future. They believe that the education system's orientation should not only consider prescribed things; instead, the development of the education system, with the help of all stakeholders involved in the education system, involved: the school leaders; the parents; the teachers and the youth groups; universities; and social partners (OECD, 2018). The framework's vision is that IT is not only about technology, but it is about the fact that every learner should develop to become a whole person. They must be able to fulfill themselves to fulfill their full potential and have a future built for the well-being of individuals, communities, and the planet (OECD, 2018).

Figure 3. OECD learning compass 2030
Source: Adapted from OECD Learning Compass (2018)

When the future of educational systems goes in the correct direction, it can make all the difference. Also, it was believed that from 2018 onwards, all children would have to know they do not have limitless resources and must also care about society. Therefore, the children would be ready for the challenges they would face and not worry about them.

There is an immediate need for changes in the educational systems because of the changes happening in society. The many challenges that exist are: (1) environmental; (2) economic; and (3) social. If we considered these one at a time, we could see the existing problems. In case of environmental concerns, there are fast depleting resources. On the other hand, many economic circumstances, scientific developments, and developed technologies help us build our lives. Yet, on the other hand, it leads to disruptive waves of change in every sector. People are still raising questions: with the development of AI, will we have any part of the system that will be human? Economic interdependencies exist at local, national, and regional levels, but with the uncertainties and risks that exist, there is not much that one can predict. The third major challenge is the social challenge of global population growth, migration, urbanization, increasing diversity, and changing communities. In many parts of the world, living standards, there are, on the other hand, many political concerns.

These concerns will affect society, and all partners must come together to address these global concerns. The OECD Education 2030 framework also contributes to the UN's 2020 Global Goals for sustainable development, so there is sustainability through partnerships for people and the planet (OECD, 2018).

The OECD also envisages that unless we consider the rapid advances in science and technology strategically, we will further widen the inequities, ensure further social fragmentation, and lead to increased speed of resource depletion. More recently, in the 21st century, there has been a high focus on well-being (OECD, 2018). However, well-being is not merely accessing material resources, income or wealth, jobs and earnings, and housing. Well-being also means many more aspects, such as quality of life, civic engagement, social connection, education, security, life satisfaction, and the environment. Further, the equity in being able to receive these would lead to inclusive growth. The education system has a major role in developing knowledge, skills, and values to provide everyone with an equitable and sustainable future. The new education systems need to prepare future generations for work environments and equip them with skills to become active and engaged citizens.

Learners navigating through the uncertainties of the future have many things to remember. First, the future generation of students needs to apply agency in their learning and their lives. This concept means they must be responsible for being part of the future world. They need to understand that things are changing so fast that they also must put in extra effort to embrace the changes and be able to influence everyone and everything around them.

Educators must understand the concept of agency and consider learner individuality and the relationships that need development with other education system partners. The idea of co-agency must be initiated while the educational systems are changing for the future (OECD, 2018). Everyone should be considered a lifelong learner, not only the students but also the educators.

Students must be prepared for the future. They need two skill-based competencies for the future, which are much more than knowledge and skills. These competencies include people's attitudes and values apart from knowledge and skills so that they also have the mindset of continually acquiring new knowledge. They should be able to apply their knowledge in changing and evolving environments. They need cognitive and meta-cognitive skills such as critical thinking, creative thinking, learning, and self-regulation; social and emotional skills such as empathy, self-efficacy, and collaboration; and para-physical skills such as new information and communication technology devices (OECD, 2018).

CONCLUSION

This chapter focused primarily on new humanizing technology-based systems, including educational systems, and exploring the global discussions to establish different organizational educational contexts. How the various stakeholders would actively participate in getting the educational systems in place for the future must be understood. It is important to understand the skills the future generation needs so that all stakeholders, namely educators, administrators, and students, can act to acquire the new skills. Many frameworks have been identified for the future, and they need to be followed carefully so that one can achieve an educational system that is sustainable for the future. Moreover, this ongoing process needs to be investigated fully so that the educational systems developed now will be sustainable for the long term. Establishing these processes will take a long time, yet we need to put in the effort.

REFERENCES

Bakkar, M. (2016). *An investigation of mobile healthcare (mHealthcare) training Design for Healthcare Employees in Jordan.* Academic Press.

Butler, D. L. (2002). Individualizing instruction in self-regulated learning. *Theory into Practice, 41*(2), 81–92. doi:10.120715430421tip4102_4

Cabinet Office. (2023). *Society 5.0.* http://www8.cao.go.jp/cstp/ english/society5_0/ index.html

Conderman, G., Ikan, P. A., & Hatcher, R. E. (2000). Student-led conferences in inclusive settings. *Intervention in School and Clinic, 36*(1), 22–26. doi:10.1177/105345120003600103

Council, B. (2018). *Future skills supporting the UAE's future workforce*. British Council. www. Britishcouncil. Ae/Sites/Default/Files/Bc_futureskills_ english_1mar18_3. pdf

Ehlers, U.-D., & Kellermann, S. A. (2019). Future skills: The future of learning and higher education. Karlsruhe.

EmiratesSkills. (2023). https://www.emiratesskills.ae

Future Skills. (2022). *Future Skills -Kingston University-UK*. https://www.kingston. ac.uk/aboutkingstonuniversity/future-skills/

Güğerçin, S. (2021). How Employees Survive In The Industry 5.0 Era: In-Demand Skills Of The Near Future. Int. J. Discip. Econ. Adm. Sci. Stud. *IDEAstudies*, 7(31), 524–533. doi:10.26728/ideas.452

Hallion, A. M. (1994). *Strategies for Developing Multi-Age Classrooms*. Academic Press.

IBE-UNESCO. (2017). *Training Tools for Curriculum Development: Personalized Learning*. https://unesdoc.unesco.org/ark:/48223/pf0000250057/PDF/250057eng. pdf.multi

IGI Global Dictionary. (2023). *What is Skill*. https://www.igi-global.com/dictionary/ combining-local-global-expertise-services/27088

Lantada, A. D. (2022). Engineering education 5.0: Strategies for a successful transformative project-based learning. *Insights Into Global Engineering Education After the Birth of Industry 5.0*, 19.

Learning Compass. (2018). *The OECD Learning Compass 2030*. OECD. https:// www.oecd.org/education/2030-project/teaching-and-learning/learning/

Lynch, M. (2018, January 2). *How to Humanize the Education Machine*. https://www. edweek.org/education/opinion-how-to-humanize-the-education-machine/2018/01

Muñoz, J. L. R., Ojeda, F. M., Jurado, D. L. A., Peña, P. F. P., Carranza, C. P. M., Berríos, H. Q., Molina, S. U., Farfan, A. R. M., Arias-Gonzáles, J. L., & Vasquez-Pauca, M. J. (2022). Systematic Review of Adaptive Learning Technology for Learning in Higher Education. *Eurasian Journal of Educational Research*, 98(98), 221–233.

OECD. (2018). *The Future of Education and Skills: Education 2030*. OECD Publishing. https://www.oecd.org/education/2030/E2030%20Position%20Paper%20 (05.04.2018).pdf

Oxford Learners Dictionaries. (2023). Skill. In *Oxford Learners Dictionaries*. https://www.oxfordlearnersdictionaries.com/definition/american_english/skill

PECB University. (2018). *ISO 21001:2018 – Educational organizations – Management systems for educational organizations – Requirements with guidance for use*. https://pecb.com/whitepaper/iso-210012018--educational-organizations--management-systems-for-educational-organizations--requirements-with-guidance-for-use

Richardson, W., & Mancabelli, R. (2011). *Personal learning networks: Using the power of connections to transform education*. Solution Tree Press.

Sands, E. G., Bakthavachalam, V., & Reddick, R. (2020). *Global skills index 2020*. Academic Press.

Saxena, A., Pant, D., Saxena, A., & Patel, C. (2020). Emergence of educators for industry 5.0: An Indological perspective. *International Journal of Innovative Technology and Exploring Engineering*, 9(12), 359–363. doi:10.35940/ijitee. L7883.1091220

SkillsUSA. (2023). www.skillsusa.org

SkillsFuture Singapore. (2022). *Skills Demand For the Future Economy*. https://www.skillsfuture.gov.sg/-/media/Skills-Report-2021/Skills-Report-Documents-FINAL/SSG-Skills_Demand_for_the_Future_Economy_2021.pdf

The Global Goals. (2015). *Goal #4 Quality Education*. https://www.globalgoals.org/goals/4-quality-education/?gclid=Cj0KCQjwiZqhBhCJARIsACHHEH-eh34l9Gu xdCPCqItBp9mNLFiEozM9zWoGHnnun9PpMZv84ARV2hgaAoQ8EALw_wcB

United Nations. (2015). *Sustainable Development Goals*. https://www.un.org/sustainabledevelopment/blog/2015/12/sustainable-development-goals-kick-off-with-start-of-new-year/

Section 5
Industry 5.0: Transforming the Healthcare Industry

Chapter 8
Regulatory Shift Healthcare Applications in Industry 5.0

Rita Komalasari
Yarsi University, Indonesia

ABSTRACT

This chapter aims to present pertinent theoretical frameworks and the most recent results of empirical research in health. It is written for amateurs and experts who wish to understand the industry better. A meta-ethnographic synthesis of research on healthcare applications for Industry 5.0 forms the basis of the review's method. The results show that the link between Industry 5.0 and healthcare applications have a positive relationship. Big data may be used by Industry 5.0 to learn new things and produce symmetrical innovation. It also establishes a digital knowledge network that offers accurate medical data and vital patient records. It establishes a digital knowledge network that offers accurate medical data and vital patient records. This chapter contributes to a better understanding of the link between Industry 5.0 and healthcare applications.

INTRODUCTION

This chapter contains reviews based on the concept of the book *Advanced Research and Real-World Applications of Industry 5.0*. Both academics and the public will benefit from this chapter. For example, a health professional is responsible for distributing healthcare resources. Regulation changes are required for industry-wide healthcare applications in 5.0 (Madikunta et al., 2022). Moreover, its capacity to develop and deliver cheap healthcare products is discussed in this chapter. The chapter's structure is as follows: A short literature-based strategy was presented as

DOI: 10.4018/978-1-7998-8805-5.ch008

a starting point. Second, we introduced the findings of the review of the literature in the section devoted to them. As a final part of the debate, the author provides an evidence-fusion theory-based method, which includes proof of the link between healthcare applications for Industry 5.0. In the concluding part, we offer a conclusion and ideas for further study.

BACKGROUND

Over the last decade, there has been an emerging interest in understanding the link between Industry 5.0 and healthcare applications. Healthcare consumers today strongly emphasize tailoring products to meet their specific needs. The term "industry 5.0" refers to a fifth industrial revolution where the tailored needs of consumers involved in the provision of health care might be met. This chapter explores regulatory changes in Healthcare applications in Industry 5.0. It can develop and deliver affordable healthcare products.

This section provides information on developing and developed countries' medical device markets and regulatory systems. As one of the world's most competitive medical device industries, the UK is renowned for its capacity to innovate medical equipment for the UK and international marketplaces consistently (Morrar et al., 2017). This is partly attributed to more excellent R & D investment levels and easier access to venture capital than the SA industry, which lacks government backing and access to venture capital—support for R & D and technical advancement (Foray et al., 2012). Despite the UK's solid domestic product development capacity and strict regulations, most of the treatment equipment supply in Southeast Asia comes from imports. There are limited regulations there, creating significant prospects for international device producers. As a result of increasing expenses related to healthcare technology R & D, clearance, compliance, and quality control in the UK and Southeast Asia are not widely accessible (World Health Organization 2013). Its purpose is to investigate how legislative changes affect industrial capacity and the creation of affordable medical equipment for regional communities in the UK and Southeast Asia, respectively.

By constantly changing health regulations, society's safety has been altered (Kushi et al. 2012). For example, medical device issues in the UK, such as hip replacements and breast implants, were essential in developing new regulations (Melvin & Torre, 2019). Because of these concerns, the Southeast Asian government implemented a series of health regulatory adjustments (Bao et al., 2020). Processes such as the use of surprise inspections of manufacturing facilities by notified bodies, improved cooperation in the oversight of notified bodies, preservation of uniformity in recognition of notified entities across EU member states, improvement of monitoring systems, and the use of medical device traceability tools, all contributed to regulatory changes.

The rising manufacturing process and rules have changed due to a demand for health technology. This governs the availability of such devices on the market (Dawoud et al., 2021). Other catalysts for the change are the shifting legal landscape around medical care delivery and the concerns expressed by patients about their experiences with poor medical treatment. As a result, there is a void in our knowledge of how legal changes impact efforts to make affordable medical technology and gadgets available in developed and developing countries. As a result, this chapter fills in the knowledge gaps left by past research. The following section gives a technique for doing a literature study on healthcare technology and medical device regulation.

METHOD

A meta-ethnographic synthesis of research on healthcare applications for Industry 5.0 forms the basis of this review's methodology. The current evaluation focuses on qualitative studies reporting on healthcare applications for Industry 5.0. Professional understanding of healthcare applications for Industry 5.0, how they affect practice, and the obstacles and facilitators for healthcare applications for Industry 5.0 mainly interested us. By combining the perspectives of experts with different roles and responsibilities, we may better understand the situation in healthcare applications' readiness for Industry 5.0. Furthermore, this data will shed light on successful techniques experts use and highlight areas where further healthcare applications for Industry 5.0 help are required.

RESULTS

The following section presents the healthcare applications applicable to Industry 5.0. To put this study in perspective with the existing body of knowledge on medical device regulation, the relevant healthcare technology regulation literature and medical device regulation specifically will be thoroughly reviewed in this chapter. In addition, the author specializes in academic studies on medical technology laws in developing and emerging nations. Therefore, this work also covered regulated industries in this literature review.

Countries With More Advanced
Regulations on Medical Devices

In advanced countries, the medical industry is moving toward personalization, which calls for a gadget to satisfy individualized needs like checking blood pressure, sugar levels, and other bodily data (Bakkar & Axmann, 2022). Since wearable technologies provide clinicians access to patients' real-time health information and integrate that information with individual medical records, individuals and groups of patients can benefit from data mining. Doctors can prescribe the proper medication following each patient's needs, thanks to the intercommunication capabilities of these clever gadgets. This revolution uses various applications to monitor our everyday routines and lives. It presents artificial intelligence that fundamentally transforms human existence and discovers how our bodies respond. It helps the medical industry innovate while automating the manufacturing process.

In a fast-changing area, legal complexities may confound the regulation of medical devices (Onwudiwe et al., 2018). We have done many studies on medical device regulation in sophisticated nations over the last twenty years, although they are few compared to those on drug regulation. This section focuses on the control and advancement of medical equipment. Many researchers have studied medical device innovation, including Guo et al. (2016) and Reyes et al. (2021). These and other studies have examined the impact of regulatory and regulatory reform on medical device innovation. Research on the influence of regulation of medical device innovation explores and emphasizes themes like resource scarcity and allocation. According to Reyes et al. (2021), there are two ways in which regulatory limitations affect innovation: To begin with, we attached various costs to regulatory changes. The second potential benefit of regulation is that it may encourage innovation by allowing businesses to expect higher profits. These authors contend that the increased profits projected might be due to either an increase in manufacturing output or a regulatory influence on the items' prospective entry into the market. According to Reyes et al. (2021), the most evident is that they prioritized compliance over product creation while using productive resources. Jnr (2020) examines the role of governance in the innovation process. According to the report, regulation needs to catch up to technological progress. However, according to Rayan et al. (2021), legislative hurdles to innovation impede the new medical gadget's introduction to the healthcare system. According to Lee et al. (2018), open innovation methods may help medical device companies control several stakeholders' thoughts and remove current barriers to market entry. Several case studies show the benefits of interdisciplinary collaboration in developing new medical equipment (Lett et al., 2006), effectively overcoming the obstacles in medical product innovation. For medical technology, Stern (2017) found that small enterprises are less likely to pioneer due to the higher costs involved in doing so for more financially limited companies because the regulatory process influences market entrance patterns. To better understand the role of regulation in the innovation process, we will look at the entire innovation cycle, which includes the allocation of resources for innovation, the development of the innovation process itself and manufacturing, and the sales and usage of the final goods.

Several studies have compared and contrasted the regulatory systems for medical devices in the EU and the US and the opportunities for the growth of these two regions (Ershova, 2011). Research has shown differences between the two systems: The United States system is highly centralized, with FDA authority over all processes for a product's admittance to the market. It has been "outsourced" to an external entity, a "notified body," per European legislation on medical devices. As a result, USA regulations have been regarded as more restrictive, and we have seen this slowing down innovation. Patients in Europe get quicker access to some technologies. However, many items come with softer performance guarantees, which might be more likely to have later-identified negative consequences (García, 2020). As a federal organization, the US regulatory authority has enormous powers. Comparative studies have found this to be true. European product certification partners are frequently smaller than businesses looking to get their products cleared (CE certified) for the European market. Because companies may move, a product can be allowed before it gets approved for sale in the USA, since the essential stages are completed more quickly in Europe. The notified body certification approach offers a lower entrance hurdle than the government-owned certification mechanism now in the USA. Comparative studies show manufacturers make device categorization decisions in both regulatory regimes. Based on the existing rules, manufacturers choose which class a product should fit into and apply for approval accordingly (Brönneke et al., 2021).

Only some medical device companies exist in undeveloped nations, and those that specialize in lower-end products. Medical devices play an increasingly important role in developing nations because of the wide range and severity of health problems. Only a few medical device makers are based in developing countries, according to WHO (2017). Almost all the medical equipment used in public hospitals in developing nations is imported, with just a tiny portion produced on-site (Diaconu et al., 2017). In addition, most medical gadgets are unsuitable for the region and can only be maintained with adequate infrastructure (WHO, 2020). For instance, a thorough evaluation of the treatment device legislation in Indonesia was conducted to identify opportunities for homegrown medical device development. Despite the research's conclusions, there needed to be more regional manufacturing capability and a mechanism to reward companies for developing high-demand medical items.

The same study results show that financing and support for research and development (R & D) must be improved for goods with the potential to improve public health.

DISCUSSION

The following section presents a discussion of healthcare applications for Industry 5.0. According to Smith et al. (2018), three major criticisms of health regulation have been outlined. The first criticism is that the existence of regulation may successfully avoid disruptive technological advancements since it raises the cost of innovation, decreasing the availability of inexpensive healthcare solutions. Disruptive innovation theory supports this claim. Regulation may stifle sector growth by erecting immovable entry hurdles that retard the sector's development, as the author concludes in his analysis. According to Tox, we should separate research, development, and manufacturing capacities. For Smith et al. (2018), a regulatory system that increasingly encompasses worldwide regulatory frameworks must be understood in collaboration with the industry's capabilities. Institutions, as well as other developments in scientific regulation. Manufacturing and innovation capabilities are equally significant in developing nations, according to Nakandala et al. (2015).

Many academics have rigorously researched how technological developments affect market structures (Treiblmaier & Sillaber, 2021). Regulation and regulatory change's impact on industrial capacity has received much less attention. Research on particular companies and their implementation efforts are scarce in the literature. In addition, no studies compare the success of various businesses' ability to adjust to new legislation. Investigating the implementation activities companies take in response to a regulation change may provide us with fresh insight into the behavior of enterprises. Several studies have examined the evolution of businesses, such as the pharmaceutical sector, from patenting to output (Khanna et al., 2016). The medical device business has received microscopic scrutiny in this area. As Sherman and colleagues (2016) found in their legislative and policy framework analysis, medical device regulation and development studies have gotten greater attention than pharmaceutical product regulation and development studies.

In this chapter, a few scholars have attempted to explore the regulation of medical devices. New legislative developments and practice changes need more study, particularly considering the ever-expanding information (Auger et al., 2021). A systematic study by Callea et al. (2022) compared the regulatory frameworks for medical devices in the European Union and the United States of America. Their findings noted that regulatory regimes must grow, but this systematic study did not guide legislators or regulators looking to alter device regulation. We will use sectoral innovation systems in this study's suggestions to policymakers developing national, sectoral, or regional innovation programs to get around this restriction (Schot et al., 2018). According to WHO (2013), there is a differing degree in the UK's capacity to manufacture medical devices and Southeast Asia. No comprehensive research has analyzed how medical device regulation affects commercial healthcare technology development from an evolutionary standpoint and barriers to their commercialization at an affordable cost. Studying this further would be worthwhile since it would help us better grasp the subject.

This chapter analyzed literature data from businesses with bases in the UK and Southeast Asia to compare incidences of regulatory change in those two countries. First, the EU system was the foundation for both countries' medical device regulation frameworks. Despite the nation's stringent rules, Southeast Asia has adopted or harmonized European criteria into its law (Teo et al., 2016). A third aspect is that the UK and Southeast Asia legal systems adhere to the IMDRF's (International Forum of Medical Device Regulators) ideology of achieving global regulatory convergence for medical device standards. As a result, the Global Harmonization Task of both study nations (GHTF implemented Force's Risk Classification System). Their findings spurred the UK to pioneer risk-based regulation, and governments have studied and emulated the UK's regulatory reform efforts worldwide (Brandon, 2012), including those in Southeast Asia (SA). SA has taken a page or two from the UK using regulatory frameworks.

A method based on the SSI method applies to assessing the current medical device business dynamics, including the components of the literature-gathered medical device manufacturing procedures. Using the SSI technique in research might enhance our ability to understand regulatory change, its reasons, and how it affects people's ability to produce inexpensive healthcare innovations. The strengthening theory will be utilized to examine the changes in legislation, industrial competence, and how these three sectors interact to generate inexpensive healthcare innovations. First described at a national level, Sectoral Innovation Systems (SSI) and National Innovation Systems (NIS) (SSI) follow in specific ways (NIS). Several writers contributed to the NIS framework, presenting the country's components differently. Some scholars (Garmann et al., 2017) claim that national manufacturing system design and public policy influenced innovation. As WHO (2017) informed, both geographically and culturally, people must be close to one another for inventive capacities to develop.

This study's assessment of the medical devices industry causes an analysis of the formal institutions or organizations engaged in this national sector's innovation. Regarding the NIS system's effectiveness, it has to include supply-side and demand-side factors and the importance of national sovereignty in an era where technology and science are produced and are more globalized (Atun et al. 2015). Besides the fact that NIS literature provides more of a conceptual framework for analyzing macro-level elements particular to a nation than a formal theory, these factors are examined at great length in this book. As a result, coexisting NIS will be considered in this study at many analytical levels. The most relevant viewpoint for the sectoral nature of the medical device business led to the Sectoral Production and Innovation System as the primary method for this study (Zhang et al., 2013). Many elements influence innovation and output in different industries, and the Framework for Sectoral Systems of Innovation (SSI) considers them. Rather than focusing on individuals or institutions, it emphasizes businesses and the skills they develop in doing business.

However, other essential aspects like the diversity of players, networks, demand, and institutions are all primary focus in the framework. Regulatory developments in the medical device industry, both intentional and unexpected, are examined in this paper and used as a case study in SSIs. Approaches like those of SSI are dynamic, adaptive, and grown environments from a process perspective because we use the SSI framework in this study, which has an evolutionary approach deemed acceptable. A system (group) of businesses develops and manufactures a sector's products. The group is also involved in the creation and use of its technologies. Creating artifact technologies via processes of interaction and collaboration and selecting enterprises through competition and selection are two ways such a system of businesses is connected. For example, the creation of new products and market activity; another description was the collection of new and used products and agents engaging within and outside the market to create, produce, and sell these things. There are five pillars to the SSI: knowledge and technology, actors (e.g., businesses), and networks, together with structures like norms, laws, rules, and regulations. SSI builds on these pillars. Therefore, considering these crucial factors in understanding how an industry's dynamics and innovation processes work, according to SSI theory, innovative new technologies are born when these elements come together correctly. Research on sectoral innovation systems (SSI) stresses the diversity of organizations and individuals involved and the variances in the knowledge base. According to some scholars, sectors and innovation systems determine how individuals and organizations interact at distinct geographical levels because of this concentration on unique sectoral features (Meuer et al., 2015).

Southeast Asia's medical device business has flourished since introducing the REDs legislation. Southeast Asia's entire value chain for medical equipment and diagnostics. These values include product manufacturers, regulators who oversee them, funders who support them, and government organizations and business associations that helped create the regulations. In 2016, new regulations on medical devices and In Vitro Diagnostic Medical Devices (IVD) were issued. In the last several years, certain medical device industry players in Asia have been campaigning for this regulation to have an impact (Chen et al., 2018). Some R&D businesses increasingly rely on local and overseas universities and research institutions. These collaborations aim to satisfy the regulatory criteria for clinical proofs to supplement internal R & D resources, access facilities and equipment, and, to a lesser extent, transfer innovative technology to businesses. Strong university links have been used.

Research and development interests are highly associated with a company's relationship with domestic universities and research institutions. Also, in line with this theory, regulators foster knowledge creation, inventiveness, and regulatory solutions through substantial cooperation or tight connections between producers and specialized markets. We expect these findings from the literature. While enterprises, universities, and government organizations are examples of non-corporate entities that engage in learning and knowledge-creation activities in sectoral innovation systems, this is only sometimes the case. Most multinational corporations (MNCs) said that the third regulation modification (REDs) impacted their product supply chain, resulting in better product quality and safety. Due to this move, they gained a competitive edge in local and international markets. As a result, radiation-emitting device regulations call for local SA businesses to import any specific electro-medical device from scratch or refurbish the equipment (Ng et al., 2021). Responses showed that most companies face significant obstacles in this area. Because of the high standards, domestic vendors in Southeast Asia found getting into the supply chain more challenging.

The link between Industry 5.0 and healthcare applications would have a positive relationship. Big data may be used by Industry 5.0 to learn new things and produce symmetrical innovation. It also establishes a digital knowledge network that offers accurate medical data and vital patient records. It creates a digital knowledge network that offers accurate medical data and vital patient records.

SOLUTIONS AND RECOMMENDATIONS

Mass customization is made possible by Industry 5.0, resulting in a more specialized medical component. These individualized elements are employed in the patient's medical care to meet their specific needs and way of life. In this revolution, accurate manufacturing of personalized medical implants, artificial organs, bodily fluids, and transplants is possible because of industry 5.0 enabling technologies. This chapter contributes to a better understanding of the link between Industry 5.0 and healthcare applications.

CONCLUSION

The ideas, conceptions, studies, and industrial capacities related to medical device regulation have been discussed in this chapter. The sectoral systems of innovation (SSI) paradigm was used to examine how regulatory changes affect industry capacity. Analyzing a transformative process is at the heart of the study of regulatory reform. Because of SSI's evolutionary foundation and its emphasis on process rather than product, they widely accepted it as a viable framework for studying coevolution. Evolutionary theory may explain regulatory developments in the medical device industry. The evolutionary theory's behavioral underpinning is based on adaptation and new findings. Technological advancement and innovation are crucial in evolutionary theory because they affect the selection process and result in systemic shifts. Therefore, technical skills are essential to this study's analytic approach. This chapter summarises that a company can only grow if it uses all of its current resources to its most significant potential, making this study essential and highly original in Industry 5.0 and concerning the Health industry specifically.

REFERENCES

Atun, R., De Andrade, L. O. M., Almeida, G., Cotlear, D., Dmytraczenko, T., Frenz, P., Garcia, P., Gómez-Dantés, O., Knaul, F. M., Muntaner, C., de Paula, J. B., Rígoli, F., Serrate, P. C.-F., & Wagstaff, A. (2015). Health-system reform and universal health coverage in Latin America. *Lancet*, *385*(9974), 1230–1247. doi:10.1016/S0140-6736(14)61646-9 PMID:25458725

Auger, S. D., Jacobs, B. M., Dobson, R., Marshall, C. R., & Noyce, A. J. (2021). Big data, machine learning, and artificial intelligence: A neurologist's guide. *Practical Neurology*, *21*(1), 4–11. PMID:32994368

Bakkar, M. N., & Axmann, M. (2022). Industry 4.0: Learning Analytics Using Artificial Intelligence and Advanced Industry Applications. In *Manage Your Own Learning Analytics* (pp. 193–204). Springer. doi:10.1007/978-3-030-86316-6_9

Bao, Y., Sun, Y., Meng, S., Shi, J., & Lu, L. (2020). 2019-Nov epidemic: Address mental health care to empower Society. *Lancet*, *395*(10224), e37–e38. doi:10.1016/S0140-6736(20)30309-3 PMID:32043982

Brandon, E. (2012). *Global approaches to site contamination law*. Springer Science & Business Media.

Brönneke, J. B., Müller, J., Mouratis, K., Hagen, J., & Stern, A. D. (2021). Regulatory, legal, and market aspects of intelligent wearables for cardiac monitoring. *Sensors (Basel)*, *21*(14), 4937. doi:10.339021144937 PMID:34300680

Callea, G., Federici, C., Freddi, R., & Tarricone, R. (2022). Recommendations for designing and implementing an Early Feasibility Studies program for medical devices in the European Union. *Expert Review of Medical Devices*.

Chen, Y. J., Chiou, C. M., Huang, Y. W., Tu, P. W., Lee, Y. C., & Chien, C. H. (2018). A comparative study of medical device regulations: US, Europe, Canada, and Taiwan. *Therapeutic Innovation & Regulatory Science*, *52*(1), 62–69. doi:10.1177/2168479017716712 PMID:29714608

Dawoud, D., Naci, H., Ciani, O., & Bujkiewicz, S. (2021). Raising the bar for using surrogate endpoints in drug regulation and health technology assessment. *BMJ (Clinical Research Ed.)*, 374. doi:10.1136/bmj.n2191 PMID:34526320

Diaconu, K., Chen, Y. F., Cummins, C., Jimenez Moyao, G., Manaseki-Holland, S., & Lilford, R. (2017). Methods for medical device and equipment procurement and prioritization within low-and middle-income countries: A systematic literature review findings. *Globalization and Health*, *13*(1), 1–16. doi:10.118612992-017-0280-2 PMID:28821280

Ershova, N. (2011). *Medical device regulations as a source of industrial leadership: A comparative study of American and European regulatory approaches*. Academic Press.

Foray, D., Mowery, D. C., & Nelson, R. R. (2012). Public R&D; and social challenges: What lessons from mission R&D; programs? *Research Policy*, *41*, 1697-1702.

García-Sánchez, I. M., & García-Sánchez, A. (2020). Corporate social responsibility during the COVID-19 pandemic. *Journal of Open Innovation*, *6*(4), 126. doi:10.3390/joitmc6040126

Garmann-Johnsen, N. F., & Eikebrokk, T. R. (2017). Dynamic capabilities in e-health innovation: Implications for policies. *Health Policy and Technology*, *6*(3), 292–301. doi:10.1016/j.hlpt.2017.02.003

Guo, X., Wang, J., Zhao, W., Zhang, K., & Wang, C. (2016). Study medical device innovation design strategy based on demand analysis and process case base. *Multimedia Tools and Applications*, *75*(22), 14351–14365. doi:10.100711042-015-3176-2

Jnr, B. A. (2020). Examining the role of green IT/IS innovation in collaborative enterprise-implications in an emerging economy. *Technology in Society*, *62*, 101301.

Khanna, R., Guler, I., & Nerkar, A. (2016). Fail often, fail big, and fail fast? Learning from small failures and R&D performance in the pharmaceutical industry. *Academy of Management Journal*, *59*(2), 436–459. doi:10.5465/amj.2013.1109

Kushi, L. H., Doyle, C., McCullough, M., Rock, C. L., Demark-Wahnefried, W., Bandera, E. V., Gapstur, S., Patel, A. V., Andrews, K., & Gansler, T. (2012). American Cancer Society Guidelines on nutrition and physical activity for cancer prevention: Reducing cancer risk with healthy food choices and physical activity. *CA: a Cancer Journal for Clinicians*, *62*(1), 30–67. doi:10.3322/caac.20140 PMID:22237782

Kwon, S. (2009). Thirty years of national health insurance in South Korea: Lessons for achieving universal health care coverage. *Health Policy and Planning*, *24*(1), 63–71. doi:10.1093/heapol/czn037 PMID:19004861

Lee, M., Yun, J. J., Pyka, A., Won, D., Kodama, F., Schiuma, G., Park, H. S., Jeon, J., Park, K. B., Jung, K. H., Yan, M.-R., Lee, S. Y., & Zhao, X. (2018). How to respond to the fourth industrial revolution or the second information technology revolution? Dynamic new combinations between technology, market, and Society through open innovation. *Journal of Open Innovation*, *4*(3), 21. doi:10.3390/joitmc4030021

Lettl, C., Herstatt, C., & Gemuenden, H. G. (2006). Users' contributions to radical innovation: Evidence from four cases in the field of medical equipment technology. *Research Management*, *36*(3), 251–272.

Maddikunta, P. K. R., Pham, Q. V., Prabadevi, B., Deepa, N., Dev, K., Gadekallu, T. R., ... Liyanage, M. (2022). Industry 5.0: A survey on enabling technologies and potential applications. *Journal of Industrial Information Integration*, *26*, 100257. doi:10.1016/j.jii.2021.100257

Melvin, T., & Torre, M. (2019). New medical device regulations: The regulator's view. *Effort Open Reviews*, *4*(6), 351–356. doi:10.1302/2058-5241.4.180061 PMID:31312522

Meuer, J., Rupietta, C., & Backes-Gellner, U. (2015). Layers of coexisting innovation systems. *Research Policy*, *44*(4), 888–910. doi:10.1016/j.respol.2015.01.013

Morrar, R., Arman, H., & Mousa, S. (2017). The fourth industrial revolution (Industry 4.0): A social innovation perspective. *Technology Innovation Management Review*, *7*(11), 12–20. doi:10.22215/timreview/1117

Nakandala, D., Turpin, T., & Djeflat, A. (2015). Parallel innovation policies to support firms with heterogeneous innovation capabilities in developing economies. *Innovation and Development*, *5*(1), 131–145. doi:10.1080/2157930X.2014.980552

Ng, K. H., Brady, Z., Ng, A. H., Soh, H. S., Chou, Y. H., & Varma, D. (2021). The status of radiation protection in medicine in the Asia-Pacific region. *Journal of Medical Imaging and Radiation Oncology*, *65*(4), 464–470. doi:10.1111/1754-9485.13165 PMID:33606359

Onwudiwe, N. C., Tenenbaum, K., Boise, B. H., Elton, J., & Manning, M. (2018). *Real World Evidence: Implications and Challenges for Medical Product Communications in an Evolving Regulatory Landscape. Food and Drug Law Institute*.

Rayan, R. A., Tsagkaris, C., & Iryna, R. B. (2021). The Internet of things for healthcare: applications, selected cases, and challenges. In *IoT in Healthcare and Ambient Assisted Living* (pp. 1–15). Springer. doi:10.1007/978-981-15-9897-5_1

Reyes, D. R., van Heeren, H., Guha, S., Herbertson, L., Tzannis, A. P., Ducrée, J., Bissig, H., & Becker, H. (2021). Accelerating innovation and commercialization through standardization of microfluidic-based medical devices. *Lab on a Chip*, *21*(1), 9–21. doi:10.1039/D0LC00963F PMID:33289737

Schot, J., & Steinmueller, W. E. (2018). Three frames for innovation policy: R&D, innovation systems, and transformative change. *Research Policy*, *47*(9), 1554–1567. doi:10.1016/j.respol.2018.08.011

Sherman, R. E., Anderson, S. A., Dal Pan, G. J., Gray, G. W., Gross, T., Hunter, N. L., LaVange, L., Marinac-Dabic, D., Marks, P. W., Robb, M. A., Shuren, J., Temple, R., Woodcock, J., Yue, L. Q., & Califf, R. M. (2016). Real-world evidence—What is it, and what can it tell us? *The New England Journal of Medicine*, *375*(23), 2293–2297. doi:10.1056/NEJMsb1609216 PMID:27959688

Smith, R. O., Scherer, M. J., Cooper, R., Bell, D., Hobbs, D. A., Pettersson, C., Seymour, N., Borg, J., Johnson, M. J., Lane, J. P., Sujatha, S., Rao, P. V. M., Obiedat, Q. M., MacLachlan, M., & Bauer, S. (2018). Assistive technology products: A position paper from the first global research, innovation, and education on assistive technology (GREAT) summit. *Disability and Rehabilitation. Assistive Technology*, *13*(5), 473–485. doi:10.1080/17483107.2018.1473895 PMID:29873268

Stern, A. D. (2017). Innovation under regulatory uncertainty: Evidence from medical technology. Journal of Public Economics, 145, 181–200.

Teo, H. S., Foerg-Wimmer, C., & Chew, P. L. M. (2016). *Medicines Regulatory Systems and Scope for Regulatory Harmonization in Southeast Asia*. Academic Press.

Treiblmaier, H., & Sillaber, C. (2021). The impact of blockchain on e-commerce: A framework for salient research topics. *Electronic Commerce Research and Applications*, *48*, 101054. doi:10.1016/j.elerap.2021.101054

World Health Organization. (2013). *Medical devices and eHealth solutions: Compendium of innovative health technologies for low-resource settings 2011-2012*. World Health Organization.

World Health Organization. (2017). *China's policies promote the local production of pharmaceutical products and protect public health*. World Health Organization.

World Health Organization. (2020). Rational use of personal protective equipment (PPE) for coronavirus disease (COVID-19): Interim guidance, 19 March 2020 (No. WHO/2019-Nov/IPC PPE_use/2020.2). World Health Organization.

Zhang, L., Lam, W., & Hu, H. (2013). A case study of leading medical device companies in China is a complex product and system, catch-up, and sectoral system of innovation. *International Journal of Technological Learning, Innovation and Development*, *6*(3), 283–302.

ADDITIONAL READING

A'isy, N. R., Ernawati, K., Komalasari, R., & Gunawan, A. (2022). Relationship between Environmental Sanity and DHF Incidence: A Systematic Review and Islamic Perspectives. *Junior Medical Journal*, *1*(4), 492–503.

Ernawati, K., Fadilah, M. R., Rachman, M. A., Nadira, C., Sartika, P. A. J., Jannah, F., & Komalasari, R. (2022). Implementasi Kebijakan Program Pengendalian Demam Berdarah Dengue di Puskesmas Kresek, Kabupaten Tangerang. *Public Health and Safety International Journal*, *2*(02), 140–145. doi:10.55642/phasij.v2i02.244

Komalasari, R. (2022). A Social Ecological Model (SEM) to Manage Methadone Programmes in Prisons. In Handbook of Research on Mathematical Modeling for Smart Healthcare Systems (pp. 374-382). IGI Global.

Komalasari, R. (2022). Pemanfaatan Kecerdasan Buatan (Ai) Dalam Telemedicine: Dari Perspektif Profesional Kesehatan. *Jurnal Kedokteran Mulawarman*, *9*(2), 72–81.

Komalasari, R. (2023). Digital twin elderly healthcare services. In *Digital Twins and Healthcare: Trends, Techniques, and Challenges*. IGI Global.

Komalasari, R. (2023). History and Legislative Changes Governing Medical Cannabis in Indonesia. In *Medical Cannabis and the Effects of Cannabinoids on Fighting Cancer, Multiple Sclerosis, Epilepsy, Parkinsons and Other Neurodegenerative Diseases*. IGI Global. doi:10.4018/978-1-6684-5652-1.ch012

Komalasari, R. (2023). Telemedicine in Pandemic Times in Indonesia: Healthcare Professional's Perspective. In N. Vajjhala & P. Eappen (Eds.), *Health Informatics and Patient Safety in Times of Crisis* (pp. 138–153). IGI Global.

Komalasari, R. (2023). The ethical consideration of using Artificial Intelligence (AI) in medicine. In *Advanced Bioinspiration Methods for Healthcare Standards, Policies, and Reform*. IGI Global.

Komalasari, R. (2023). Designing Health Systems for Better, Faster, and Less Expensive Treatment. In *Exploring the Convergence of Computer and Medical Science Through Cloud Healthcare*. IGI Global.

Komalasari, R. (2023). Postnatal mental distress, exploring the experiences of mothers navigating the health care system. In *Perspectives and Considerations on Navigating the Mental Healthcare System*. IGI Global. doi:10.4018/978-1-6684-5049-9.ch007

Komalasari, R. (2023). Treatment of menstrual discomfort in young women and a cognitive behavior therapy (CBT) program. In *Perspectives on Coping Strategies for Menstrual and Premenstrual Distress*. IGI Global. doi:10.4018/978-1-6684-5088-8.ch011

Komalasari, R. (2023). *Manfaat Positif Allium Sativum L. (Bawang Putih) Dalam Kaitannya Dengan Berbagai Penyakit. Jurnal Farmasimed*. JFM.

Komalasari, R. (in press). Pelatihan Kesehatan dan Pengembangan Profesional Berkelanjutan untuk Pendidik Olah Raga. *Jurnal Ilmiah STOK Bina Guna Medan*.

Komalasari, R. (in press). Persepsi Hakim tentang Rehabilitasi Pengguna Narkoba:Tantangan dan Peluang. *Arena Hukum*.

Komalasari, R., & Mustafa, C. (2021). Meningkatkan Pelayanan Administrasi Publik di Indonesia. *PaKMas: Jurnal Pengabdian Kepada Masyarakat*, *1*(1), 20–27. doi:10.54259/pakmas.v1i1.29

Komalasari, R., & Mustafa, C. (2021). Pendidikan Profesi dan Pengabdian Masyarakat di Indonesia. *PaKMas: Jurnal Pengabdian Kepada Masyarakat*, *1*(1), 28–36. doi:10.54259/pakmas.v1i1.30

Komalasari, R., & Mustafa, C. (2022). Empowerment of Women with Narcotic Cases. *Buana Gender: Jurnal Studi Gender dan Anak*, *7*(1).

Komalasari, R., Nurhayati, N., & Mustafa, C. (2022). Insider/Outsider Issues: Reflections on Qualitative Research. *Qualitative Report*, *27*(3), 744–751. doi:10.46743/2160-3715/2022.5259

Komalasari, R., Nurhayati, N., & Mustafa, C. (2022). Enhancing the Online Learning Environment for Medical Education: Lessons From COVID-19. In Policies and procedures for the implementation of safe and healthy educational environments: Post-COVID-19 perspectives (pp. 138-154). IGI Global.

Komalasari, R., Nurhayati, N., & Mustafa, C. (2022). Kebijakan Penanganan Penyintas HIV/AIDS Di Lembaga Pemasyarakatan. *Jurnal Kesehatan Kartika*, *17*(1), 19–27.

Komalasari, R., Nurhayati, N., & Mustafa, C. (2022). Professional Education and Training in Indonesia. In *Public Affairs Education and Training in the 21st Century* (pp. 125–138). IGI Global. doi:10.4018/978-1-7998-8243-5.ch008

Komalasari, R., Wilson, S., & Haw, S. (2021). A systematic review of qualitative evidence on barriers to and facilitators of the implementation of opioid agonist treatment (OAT) programmes in prisons. *The International Journal on Drug Policy*, *87*, 102978. doi:10.1016/j.drugpo.2020.102978 PMID:33129135

Komalasari, R., Wilson, S., & Haw, S. (2021). A social ecological model (SEM) to exploring barriers of and facilitators to the implementation of opioid agonist treatment (OAT) programmes in prisons. *International Journal of Prisoner Health*, *17*(4), 477–496. doi:10.1108/IJPH-04-2020-0020

Komalasari, R., Wilson, S., Nasir, S., & Haw, S. (2020). Multiple burdens of stigma for prisoners participating in Opioid Antagonist Treatment (OAT) programmes in Indonesian prisons: A qualitative study. *International Journal of Prisoner Health*.

Mustafa, C. (2021). The Challenges to Improving Public Services and Judicial Operations: A unique balance between pursuing justice and public service in Indonesia. In Handbook of research on global challenges for improving public services and government operations (pp. 117-132). IGI Global. doi:10.4018/978-1-7998-4978-0.ch007

Mustafa, C. (2021). Key finding: result of a qualitative study of judicial perspectives on the sentencing of minor drug offenders in Indonesia: Structural inequality. *Qualitative Report*, *26*(5), 1678–1692. doi:10.46743/2160-3715/2021.4436

Mustafa, C. (2021). The view of judicial activism and public legitimacy. *Crime, Law, and Social Change*, *76*(1), 23–34. doi:10.100710611-021-09955-0

Mustafa, C. (2021). Qualitative method used in researching the judiciary: Quality assurance steps to enhance the validity and reliability of the findings. *Qualitative Report*, *26*(1), 176–186. doi:10.46743/2160-3715/2021.4319

Mustafa, C. (2021). The News Media Representation of Acts of Mass Violence in Indonesia. In Mitigating Mass Violence and Managing Threats in Contemporary Society (pp. 127-140). IGI Global. doi:10.4018/978-1-7998-4957-5.ch008

Mustafa, C., Malloch, M., & Hamilton Smith, N. (2020). Judicial perspectives on the sentencing of minor drug offenders in Indonesia: Discretionary practice and compassionate approaches. *Crime, Law, and Social Change*, *74*(3), 297–313. doi:10.100710611-020-09896-0

Suhariyanto, B., & Mustafa, C. (2022). Analysis And Evaluation Of Legal Aid In The Indonesian Court. *Jurnal Hukum dan Peradilan*, *11*(2), 176-194.

Suhariyanto, B., Mustafa, C., & Santoso, T. (2021). Liability incorporate between n transnational corruption cases Indonesia and the United States of America. *Journal of Legal, Ethical and Regulatory Issues*, *24*(3).

KEY TERMS AND DEFINITIONS

Industry 5.0: This term refers to a fifth industrial revolution where the tailored needs of consumers involved in the provision of health care might be met.

Sectoral Systems of Innovation: A comprehensive, interconnected, and dynamic picture of sectors is provided by the idea of a sectoral system of innovation and production. A collection of products and a group of actors engaging in market and non-market activities to produce and sell those items constitute a sectoral system.

Section 6
Industry 5.0: Optimising Industry 4.0 Technologies

Chapter 9

Industry 5.0 and the Collaborative Approach of Internet of Things With Artificial Intelligence

Sunil Gupta
https://orcid.org/0000-0003-3862-1980
UPES University, Dehradun, India

Monit Kapoor
https://orcid.org/0000-0002-9036-6115
Chitkara University Institute of Engineering and Technology, Chitkara University, Rajpura, India

Hitesh Kumar Sharma
UPES University, Dehradun, India

ABSTRACT

Industry 5.0 with the internet of things (IoT) gives a human touch to Industry 4.0 for the development to provide efficiency with automation using robots and machines. The advancement in artificial intelligence makes robots with attached brains like human minds. This brain-machine interface introduces the concept of Industry 5.0, where robots with IoT features tangle with humans and try to work as an agent instead of participants. This digital transformation provides opportunities and challenges. This chapter provides development by various IoT and artificial intelligence-based industries and researchers for the use of Industry 5.0 with their application. In addition, the chapter describes how IoT robots and human values collaborate to give input to Industry 5.0. Finally, the impact of Industrial 5.0 is deliberated based on the economy and manufacturing process with increased productivity.

DOI: 10.4018/978-1-7998-8805-5.ch009

INTRODUCTION: DENOTATION OF INDUSTRY 5.0

Industry 5.0 deals with personalization that allows consumers to customize their products. The Industrial Revolution 4.0 focused on digitalization, which introduced the concept of interconnected devices using the Internet of Things (IoT), artificial intelligence, and data analytics for future automation. The history of the revolution is shown in Table 1. The First Industrial Revolution was based on mechanization, which was based on mechanical production with the help of water and steam power. This steam engine replaces animals and human power. The First Industrial Revolution mainly happened in Europe and North America (Paschek et al., 2019).

The Second Industrial Revolution provides electrification through the production of electricity and engines. It enhances production by enabling electricity into the manufacturing process. The third industrial revolution occurred during the 1980s and was famous for globalization and automation; it gave birth to the Internet and computers. This automation provides a new concept in the field of information technology with the mass production of electronic equipment. The fourth industrial revolution is the complete digitalization of the 21st century and gives the concept of artificial intelligence (AI), the Internet of things (IoT), and cryptosystems with data analytics.

The fifth industrial revolution is based on the personalization and cooperation of people and machinery to customize people's things based on their needs. This revolution puts humans back into industrial production by collaborating with artificial intelligence machines like robots (Bernick, 2017). This helps increase production, customization, and personalization for existing customers. Industry 5.0 will create high-value jobs for humans. Humans will be responsible for designing and customizing the product depending on their customer's needs. The machines help in the manufacturing and production of the same. With Industry 5.0, we can automate the manufacturing process of the industry. Data analytics may help provide real-time data from machines in any field. Analytics helps to make a design better and remove the failure of any product; it improves the quality of the process and product outcome. Industry 5.0 increased the human role in designing a product with more automation and providing better information. Industry 5.0 focuses on delivering customer satisfaction with the help of hyper customization. Technologies like Blockchain, IoT, Drones, Exoskeleton, and 5G enable the sector to access Industry 5.0 in a new era of experience (Schwab, 2017). Having all of these technologies implemented per the Sustainable Development Goals (SDG) will lead our societies to evolve into society 5.0.

Table 1. History of Industry 5.0

Revolution	Year	Technology	Features
Industry 1.0	1780-1870	Mechanization	**Steam Engines:** Water and steam act as a power for mechanical production
Industry 2.0	1871-1970	Electrification	**Combustion Engines:** Electricity is used for mass production of the product
Industry 3.0	1971-1980	Automation and Globalization	**Computer and Internet:** Production through IT Systems
Industry 4.0	1981- Till	Digitalization	**Artificial Intelligence, IoT, Data Analytics, Robotics:** Digital Era and Virtual Reality
Industry 5.0	The Future	Personalization and customization	**Cooperation of Machines With People:** Innovation for deep learning, better customization, and personalization

Impact of Industry 5.0 on People and Manufacturing

Industrial 5.0 is intelligent manufacturing that revolutionizes the production process. First, it increases the efficiency of the product three times to be produced. It increases flexibility, improves the machine and sensor's performance, and reduces maintenance costs. It optimizes the logistic cost of stock, capacity, and delivery, which intends more accurate delivery and better material handling. Finally, it maintains high material quality and satisfies the customer's needs. Figure 1 shows how Industrial 5.0 plays an intelligent role in manufacturing.

Figure 1. Industrial 5.0 manufacturing process

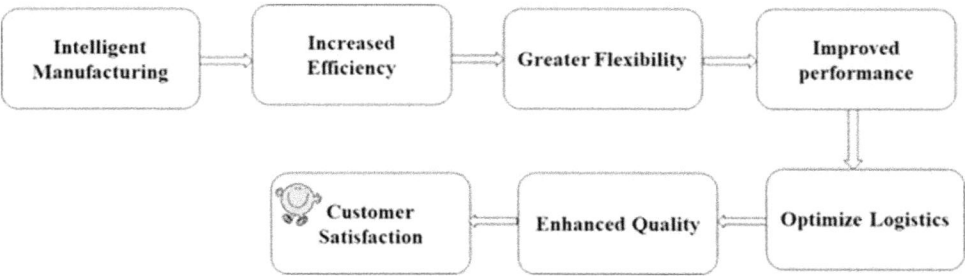

Industry 5.0 focuses on humans for manufacturing, impacting reduced waste to zero waste in manufacturing production. In addition, Industry 5.0 gives the future of collaborative robots called co-robots (Domohoske, 2019). Co-robots share a workspace

with humans, make manufacturing easier with the automatic use of machines, and provide many benefits for various applications. Figure 2 shows the collaborative robots that shake hands with humans to enhance operations and production with performance and safety (Goodrich & Schultz, 2008).

Figure 2. Co-robots: Collaborative robots act as a workforce for the future

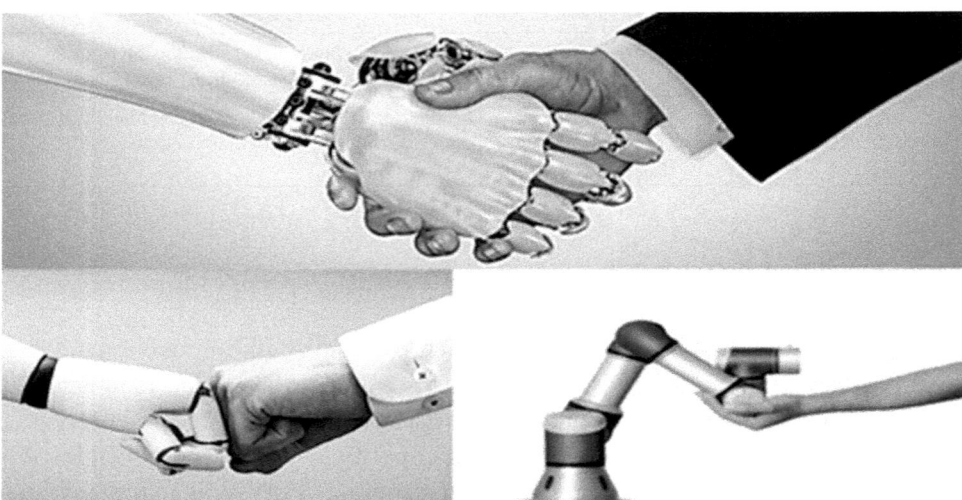

The co-robot is different from the industrial robot. They are smaller and more easily deployable and work with humans. In addition, co-robots cost less than industrial robots, which operate continuously with an efficient workforce. Some of the differences are shown in Table 2 about co-robots to industrial robots. Co-robot acts as a workforce for the future and transforms our work.

Table 2. Industrial robot and co-robot

Industrial Robot	Co-Robots
Difficult to Program: Need specified skills	Easy to Program
Speed is High	Human Speed
Large Investment required	Economical
Separate from People for safety reasons	Working with People
Heavy	Light in Weight
Work in Place of Human Employee	Working with a Human Employee

Collaboration of IoT and Artificial Intelligence in Industrial 5.0

One of the most important pillars of Industrial 5.0 is the IoT, which uses cloud technology to store, analyze and optimize the data for automation. IoT connects different devices and acts as a game-changer in Industrial 5.0 automation. It helps to create new technology, solve problems, and improve the manufacturing process. IoT helps minimize inventory costs and mass customization and improves safety in the workplace. IoT devices help the manufacturing industry automatically monitor and manage warehouses and machines (Rossi, 2018). It allows us to manage the supply chain to track inventories and estimate available resources. It helps with the removal of manual operations of documents and provides Enterprise resource plan (ERP) solutions to the company. IoT acts as a self-dependent system by using machine learning to detect and fix the issue in the machine. It makes machines automatic to heal and automate if any downtime occurs. The IoT technology helps access and control the product's manufacturing process from start to end of the production system.

Artificial intelligence plays an important role in computer vision, such as pattern recognition, face recognition, and object detection. So that systems can learn from experience by training the data and evaluating it for a certain level of accuracy. Industry 5.0 uses deep learning and machine learning to optimize the Industrial process and provide innovative solutions to the company using the data analytics process for better decision-making and optimization of the available resources. Industry 5.0 includes all emerging technologies like artificial intelligence, IoT, big data, cloud computing, mobile technologies, and many more. Figure 3 shows various pillars of Industrial 5.0, like cloud computing, artificial intelligence, data analytics, big data, mobile technology, IoT cybersecurity, and robotics (Schuster, 2020). Figure 3 shows the technologies that play an important role in collaborating different technologies with human efforts. We need human intelligence and collaboration for better results and personalization with detailed customization for having a human at the center of the manufacturing industry. Simply put, humans are better at understanding culture and different human needs.

Benefits and Opportunities for Industry 5.0

Industry 5.0 with IoT and artificial intelligence provides many benefits to industrial organizations: increased connectivity, enhanced efficiency, improved scalability, reduced downtime, better costs, and more. Figure 4 shows some major benefits; their opportunities are discussed next:

Figure 3. Pillars of Industrial 5.0: Play a role in collaboration with each other

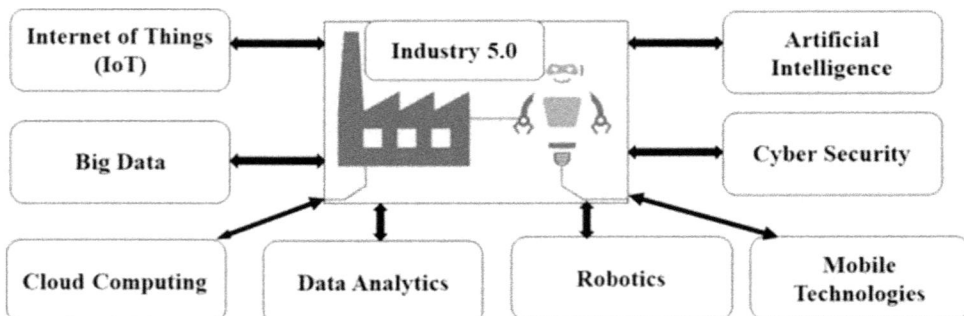

1. **Better Roles for Humans:** Industry 5.0 uses robots for labor repetitive intensive work, and it uses humans for customizing the product. It helps to enhance the customer experience to meet the customer's demand. The collaborative role of humans and robots provides a better opportunity to complete the task with better ability and critical thinking, leading to enhanced productivity (Miller, 2017);

2. **Enhanced Energy Efficiency:** Energy consumption is one of the major cost areas of any manufacturing unit. A large amount is spent on consuming energy. Therefore it becomes important to know the places where energy consumption is much more than normal. These places can be fixed and thereby save huge costs. Industry 5.0 has made it possible to measure the energy consumption of each machine and the device level (Stergaard, 2016; Özdemir & Hekim, 2018). Each device, whether overperforming or underperforming, can be tracked, enabling managers to get an insight into every piece of machinery, viz., off-hours consumption, standby–mode consumption, production capacity, and operational inefficiencies. Based on these inputs, managers can plan to reduce energy consumption, achieve higher throughput from the equipment, and find solutions for the underperforming machines. He can also plan optimized production schedules based on potential failures, optimum plans for machine operations, etc.;

3. **Predictive Maintenance and Reduced Downtime:** Industry 5.0 analyses real-time data from the sensors directly instead of historical data stored in databases and helps proactively prepare equipment maintenance. Manufacturing equipment is susceptible to wear and tear and vulnerable to certain circumstances. Schedules based on real-time data help maintain machines at the exact location at the precise moment. In addition, if any replacement of parts or repair is not needed, these parts can be utilized elsewhere, saving both time and money. For example, Industry 5.0 sensors may measure vibration and other parameters

that could lead to comparatively less than optimal operating conditions. If the vibration increases beyond a certain threshold, an alert is generated, and preventive action can be taken before the production gets affected and lessen any related downtime costs. Downtime has many other linked costs, such as lost cost when the production rate decreases to zero and loses cost of opportunity, as the company misses the opportunity to establish its goodwill in the market. Therefore, advanced techniques based on Machine Learning are also being employed to further improve today for analyzing sensor data and issuing timely advice to solve potential issues.

Figure 4. Benefits of Industry 5.0

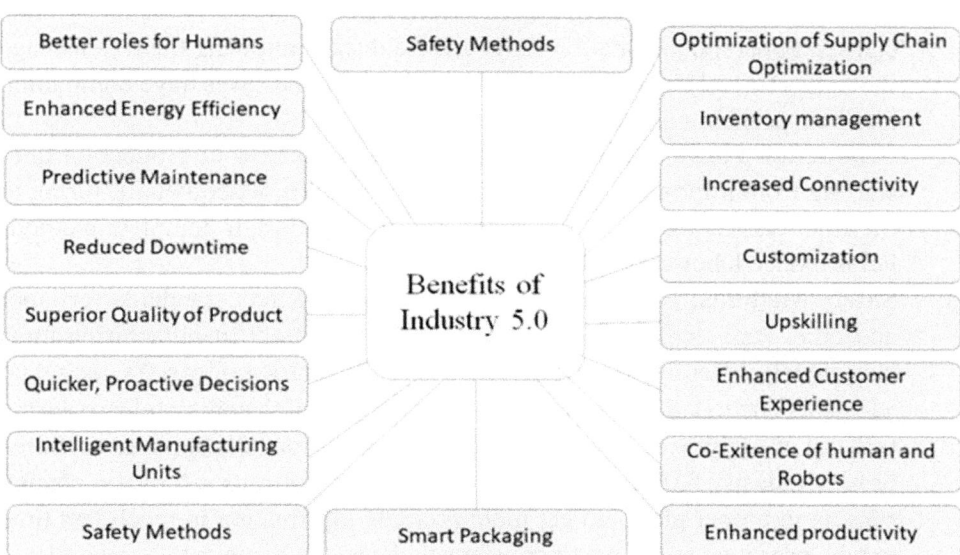

4. **Superior Quality of Product:** Improving the quality of products has many benefits, even though it increases the cost of production. Benefits include higher customer satisfaction, lower repair and maintenance costs, better market reputation, product recalls, and increased sales. One of the major reasons for producing poor-quality products is the upkeep of manufacturing equipment and their calibration. Industry 5.0 is important in helping us achieve this by embedding sensors into the equipment. Industry 5.0 sensors collect the product from various product life cycle stages. This data may include the composition of raw materials used, pressure, temperature and other environmental parameters, scrap and wastages, and the effect of paint, transportation, etc., on the final

products. The instant any sensor detects any faulty behavior of any part of the manufacturing equipment or any other parameters as mentioned above, it raises an alert that will lead to a halt in production, and the problem is rectified immediately before the quality of the product starts getting affected. Apart from the production phase, this technology helps in the development design and testing phases.

5. **Quicker, Proactive Decisions:** With Industry 5.0 in place, the manufacturing company's staff and management are fully aware of both the performance and snags of their equipment. Furthermore, as they now have access to accurate data every moment, they can act proactively, enabling them to make much quicker and more informed decisions at any instant, as per the status and requirement of the manufacturing unit.

6. **Intelligent Manufacturing Units:** Industry 5.0-enabled manufacturing units can send operational data to OEMs and field staff, allowing them to manage the units remotely. This leads to better efficiency and saves huge commuting and transportation costs.

7. **Production Flow:** Industry 5.0 helps optimize the flow of production lines starting from improving every process involved in product manufacturing to the final packaging of the product. This streamlining of complete methods helps reduce labor costs and product wastage.

8. **Safety Methods:** Better health and occupational safety standards for labor and other staff deployed using Industry 5.0. Different manufacturing units have hugely different operational practices and specific safety risks, including accidents due to falls, electricity, chemical fumes, high energy release during the start of equipment, improper ventilation, corrosion, oil slippage, etc. Some of how Industry 5.0 helps in making manufacturing units safer is by placing sensors in proper places to get more accurate information in much less time to take proactive measures to prevent accidents and damage to human life.

9. **Smart Packaging:** This is the process of enclosing manufactured products for further distribution, storage, or end-use. Industry 5.0 sensors are placed in products or packaging materials so that manufacturers can gain invaluable insights into the usage patterns and handling of products at various locations by various customers. These sensors help to track the product and analyze its deterioration, if any, during transportation. The impact of weather, road, temperature, and other environmental parameters can also be studied on the products in transit. These studies help optimize packaging solutions for better product life, reduced packaging costs, and improved customer experience.

10. **Optimization of Supply Chain Optimization:** The complete flow of goods and services (referred to as supply chain) between the organization and the customer can be effectively optimized using Industry 5.0. This includes

manufacturers, suppliers, and retailers distributing goods or services to the end customer. Industry 5.0, in this way, provides real-time connectivity amongst all the supply chain stakeholders, thereby keeping track of the complete flow of raw material, inventory, manufacturing times, distribution activities, and so on. As a result, both efficiency and pain points can be found, and corrective measures can be taken accordingly.

11.   **Inventory Management:** Industry 5.0 solutions are extensively used to manage stocks of both consumable and consumable items. Industry 5.0 permits an accurate estimate of available raw materials and supplies, enabling streamlined production flow and timely deliveries.

12.   **Increased Connectivity:** it allows all of the employees, data, and processes of the manufacturing unit/workshop to be connected to the employees, data, and methods of all the offices of executives and top management. Top management and senior executives get a complete and accurate picture of their organization and make much more informed decisions.

Challenges of the New Revolution Industry 5.0

Even though Industry 5.0 has many benefits, it has many challenges, such as developing new skills for human collaboration with: robots; soft skills; technical skills; energy efficiency; interoperability; security; and data privacy. Outside of this, interoperability and security probably pose two major challenges. Interoperability between smart objects becomes an important issue as these objects operate in various industrial and manufacturing settings, have different architectures, and deploy other protocols. Moreover, ensuring interoperability becomes difficult as these protocols and architectures need to be standardized. In addition, each organization wants its data to be secure. The explosion of sensors and other smart objects communicated to each other via not-so-secure and properly encrypted networks has given rise to security vulnerabilities. Another deterrent to implementing Industry 5.0 solutions is the cost. Many manufacturing companies find the cost of implementing Industry 5.0 to be quite high compared to ROI (return on Investment) because of the time and effort in the initial phase. Co-robots are also not that cheap, and training requires some specific cost. Many companies need upgrading technology as per Industry 5.0 (Schwab, 2017). Figure 5 shows some of the challenges in establishing Industry 5.0.

The first challenge for industry 5.0 Era implementation is related to the need for robot manufacturers' technology itself, and to acquire it needs time and effort, which is also putting pressure on people to develop their working skills needed, and that needs educational technology and systems to be further adapted with the new changes. In addition, machine-machine smart communication and decision-making need real-time data, which could be analyzed and visualized for better decisions, and it will be used for further customization of the products which will lead to more demands and sales among the global market and the supply-chain channels. Having machines and robots working together with humans is also a crucial challenge which also raises a concern for data security as well as the working environment safety, having new data security standards and frameworks in the industry 5.0 era is a crucial need to solve this challenge. Having more demand among all industries will require upgrading the old production line of manufacturers to cope with the new industry 5.0 technologies and production environment, which will be facilitated if more investment is set for this purpose.

Figure 5. Challenges of Industry 5.0

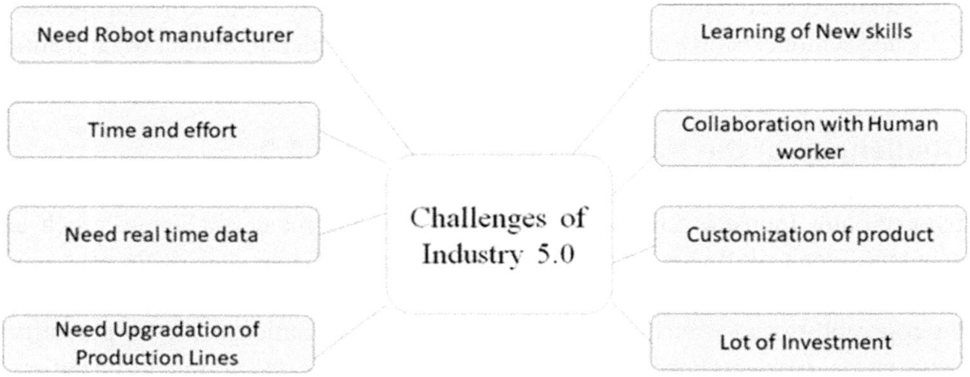

CONCLUSION

Industry 5.0 is a revolution that combines humans and machines to make manufacturing and their jobs easier and faster. This helps with the personalization and own customization of the product. It provides effectiveness in building a system dynamically. Industry 5.0 is the optimal collaboration of cloud computing, artificial intelligence, data analytics, big data, mobile technology, IoT cybersecurity, and robotics. This collaboration improves production and provides opportunities to customize the product for the consumer. This technology acts as a pillar of Industry 5.0. The chapter also discussed the various benefits, opportunities, and challenges of Industry 5.0. These futuristic technologies provide a great impact and enhanced potential for collaboration between humans and expert systems to make technology smarter and improve manufacturing. Industry 5.0 will increase demand by increasing the personalization of the products, and so, will reduce the impact of Industry 4.0 on the job market by increasing the employment rate. Industry 5.0 will enable our societies to further evolve into society 5.0, which considers the SDGs as a major guideline in its development.

REFERENCES

Bernick, M. (n.d.). *After Robots Take Over Our Jobs, Then What?* https://www.forbes.com/sites/michaelbernick/2017/04/11/after-robots-take-over-our-jobs-then-what

Domonoske, C. (2019, Apr. 30). *Even in the Robot Age, Manufacturers Need the Human Touch.* NPR.

Goodrich, M. A., & Schultz, A. C. (2008). Human-robot interaction: A survey. *Found. Trends Hum. Comput. Interact, 1*, 203–275. https://www.raconteur.net/technology/manufacturing-gets-personal-industry-5-0 doi:10.1089/omi.2017.0194

Miller, C. C. (n.d.). *Evidence That Robots Are Winning the Race for American Jobs.* https://www.nytimes.com/2017/03/28/upshot/evidence-that-robots-are-winning-the-race-for-american-jobs.html

Özdemir, V., & Hekim, N. (2018, January). Birth of Industry 5.0: Making Sense of Big Data with Artificial Intelligence, "The Internet of Things" and Next-Generation Technology Policy. *OMICS: A Journal of Integrative Biology, 22*(1), 65–76. doi:10.1089/omi.2017.0194 PMID:29293405

Paschek, Mocan, & Draghici. (2019). *Industry 5.0 – The expected impact of the next industrial revolution.* https://www.researchgate.net/publication/336653504_The_Next_Industrial_Revolution_Industry_50_and_Discussions_on_Industry_40

Rossi, B. (2018). *Manufacturing Gets Personal in Industry 5.0.* Reconteur.

Sachsenmeier, P. (2016). Industry 5.0—The Relevance and Implications of Bionics and Synthetic Biology. *Engineering (Beijing), 2*(2), 225–229. doi:10.1016/J.ENG.2016.02.015

Schuster, S. (2020, May 5). *Robotics and craftsmanship complement each other.* KUKA Blog.

Schwab. (2017). *The Fourth Industrial Revolution.* Crown Business.

Stergaard, E. H. (2016). *Industry 5.0 – Return of the human touch.* https://blog.universal-robots.com/industry-50-return-of-the-human-touch

Chapter 10
Internet of Unmanned Aerial Vehicle (IOU) in Industry 5.0

G. Prasad
Chandigarh University, India

ABSTRACT

The internet of unmanned aerial vehicles (IOU) is a layered network control architecture that is designed primarily for coordinating unmanned aerial vehicle access to controlled airspace and providing navigation with the latest innovative technology upgrades. Human-robot co-working is an emerging subject in Industry 5.0 visions. The IOU provides generic services for a wide range of such as package delivery, traffic monitoring, search and rescue, and multiple applications. In this chapter, the authors present a conceptual model of how such an architecture can be developed as well as the components of an internet of drones system based on artificial intelligence, machine learning, and digital twin are discussed. The future of drones will focus on the thrust area of computer-based domains.

INTRODUCTION

Industry 5.0 uses Industry 4.0 technologies with an optimizing approach for dealing with multiple manufacturing applications. This new industrial revolution will be able to meet the precise and sophisticated needs for high-quality components. The fifth industrial revolution is known as Industry 5.0. It consists of wirelessly connected novel technologies to improve manufacturing and automation. Industry 5.0 is a new technological innovation that enhances the interaction between humans and machines. It introduced cutting-edge technology for producing goods and services that are environmentally friendly, efficient, and safe. The concept of Industry 5.0

DOI: 10.4018/978-1-7998-8805-5.ch010

is gaining traction as intelligent and creative technologies emerge and is coming to solve the negative impact of Industry 4.0 technologies on the job market.

The Internet of Everything (IOE) is a technology that expands on the Internet of Things concept (IoT). In the context of IoT, it refers to a more complex domain that includes people and processes, as opposed to the machine-to-machine concept. This technology provides a platform for all areas of engineering and technology.

Kevin Ashton of the Radio Frequency Identification (RFID) enhancement network proposed the Internet of Things (IoT) concept in 1999. It has recently become more grounded in the world due to the proliferation of cell phones, digital transformation and cross-platform correspondence, distributed computing, and data analytics. In the field of information technology, IoT has become a well-known term. IoT transforms any real-world object into a bright object (Alipour et al., 2019). IoT gives us access to the things around us and keeps us updated on their progress.

Investigation framework that conveys innovative administrative frameworks arranged and intelligent reasoning innovation (Chen et al., 2019). The IoT system and action plans. Several devices are linked through a system provider in more innovative business models, and clients can use remarkable and supportive apps by using the right stage. They also employ cutting-edge technology and empower solid innovation to improve data collection and information activities, among other things (Ezuma et al., 2019). The Web of Things is a collection of interconnected and collaborating objects that work together to create innovative solutions. These items can be virtual real-world data or physical models. Artificial intelligence (AI) is a tool that can help track data and provide information on a real-time basis. Furthermore, unmanned flying machines used for various purposes, including surveillance integration with AI, lead to innovative technology applications.

For instance, Prasad et al. (2018) conducted detailed research on drones' influence on unmanned aerial vehicles in medical product transport. Yet, it was Prasad (2020) who analyzed the performance estimation of twin propellers in unmanned aerial vehicles. However, Prasad et al. (2018) proposed positioning UAVs using an algorithm for monitoring the forest fire region. Naveen Kumar et al. (2018) studied forest fire detection using unmanned aerial vehicles; Rajasekar et al. (2018) conducted an experimental investigation of the bio-inspired tubercle in Propeller's unmanned aerial vehicle.

BACKGROUND

Thanks to Industry 5.0 technologies, all machines and medical devices used in healthcare can now be connected. As a result, it is possible to improve machine-to-machine communication, which aids in the efficiency of the treatment process. In addition, robots, while reducing the workload of healthcare workers, improve their safety and lower human stresses at work.

Examples include electrical devices, sensors, and other physical objects. Virtual objects like sound substances and Twitter accounts can be stored, handled, and accessed. Figure 1 depicts an overview of the IoT. The Web of Things assists the core leadership process by producing, collecting, and managing critical data. The Web of Things connects a wide range of brilliant products by allowing objects (things) not originally designed to communicate and collaborate intelligently.

Figure 1. The internet of things
Source: Google Images

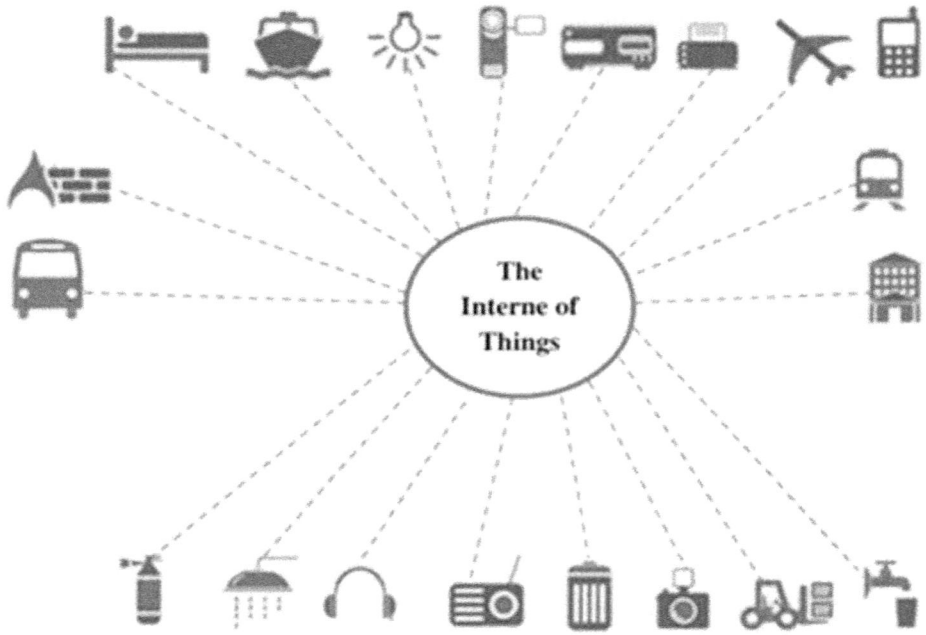

The IoT connects physical and virtual objects through interoperable data and communication technologies to enable better administration in the digital society. The IoT is a system framework that includes all its interconnected devices providing a foundation that uses interoperable communication protocols, allowing self-developing technology. This collaborative technology enables IoT devices to have personalities, physical characteristics, and virtual identities. Intelligent interfaces also support the IoT and seamlessly integrate into data manipulation and governance. According to the Global System for Mobile Communications, the IoT is "associated with life," because the revolutionary network uses intelligence to connect devices. Notions of connected life lead to personal and professional lives because they improve our healthcare, travel, training, and energy use (Ge et al., 2019).

Recent advances in machine learning algorithms, sensors, and information technology have increased the use of crewless aerial vehicles in various industries. Despite the increasing importance of data communication and electronics, computers, wireless networks, smart cities, defense, farming, and mining continue to be the core sectors—a wireless network integrated into several research projects by incorporating drones. Deep convolutional neural networks are used in UAV-based networks to understand the phenomenon of unmanned aerial vehicles (UAVs) in various environmental conditions, mainly controlled remotely. Liquid State Machines (LSM) is a machine learning model that helps to manage resources because of greater precision in decision-making and performance in the face of data variance than other learning algorithms. Sustainable, intelligent cities and military bases - unmanned aerial vehicles (UAVs) are increasingly in smart cities and the military for multiple applications. The graffiti cleaning system of the UAV system uses machine techniques and algorithms. Machine learning and system development of distinguishing between piloted and unpiloted flying objects. UAV controller signals are controlled to locate them. The controller accomplishes this by sending out RF signals. Diverse UAV types were the Wi-Fi signals detected between a UAV and its controller combined using Markov, CNN, and Naive Bayes classifiers (Khosravi et al., 2018).

AI and ML predict irrigation water requirements and manage water supplies using drones in farming. Further, it is possible through deep learning and drone photos to make agriculture sustainable. The models include Wild Jungle, Logistic Regression, and Multiple Regression for predicting and classifying vegetation production and crops. According to some sources, the following data types: wildlife, geology, and mining statistics are high-end applications. UAV-enabled machine learning is used to detect animals (Kwak et al., 2019). Surface feature detection was used to classify geotechnical mining models using machine-learning techniques such as the classification algorithm, k-nearest neighbor, random forest, Gradient Tree Booster, and Classification Relevance Vector Machine (Liu et al., 2018).

INTERNET OF THE DRONE

Because of the fulminatory expansion in many applications, the portion of Unmanned Aerial System that spreads learning has grown in recent years. The Internet of

Figure 2. The drone internet

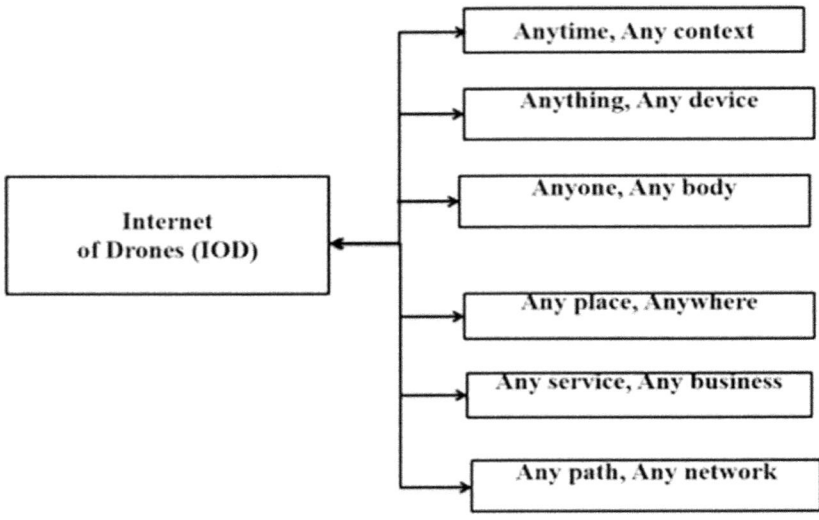

Drones image overview depicted in Figure 2. As a result, the clever Unmanned Aerial System can be designed and developed.

It combines various data and correspondence developments to control and adjusts processes in response to the needs of various clients. Improved by regularly observing and dissecting the state and exercises. The Internet of Things aspires to provide incredible variety in daily life, personal integrity, and societal profit. Because of a widely distributed, locally intelligent system of innovative inquiries, the IoT has the potential to enable extensions and modifications to essential utilities in a variety of domains while also presenting a new biological community for developing applications. Incorporating the Internet of Things concept into any vehicle would alter the nature of the journey in terms of security and traveler comfort. The fundamental concepts, definitions, characteristics, innovation, and challenges associated with the IoT (Mandloi et al.,2019) are explained. The IoT's role in developing a more brilliant Unmanned Aerial System and the importance of making compelling and convincing decisions in today's transportation. Unmanned aerial vehicles (UAVs) have grown in popularity recently, with applications ranging from photography to road transportation, inspection, and interchanges. UAVs have traditionally operated from the ground, with viewable pathway flag transmission preferred, and job go is limited, particularly in urban areas (Peng et al., 2019). The UAV has several advantages over IoT, including deployment in remote areas, transport payloads, re-programmability while in operation, and detecting almost everything. The difficulties in incorporating

UAV capabilities into training are numerous and significant (Rey et al., 2016). It is required for low-inactivity control flagging, high-precision continuous route and monitoring, rapid system topology change, and rapid media leaks. In this special issue, we intend to investigate hypothetical and research results on each aspect of UAVs over IoT. Fascination areas include, but are not limited to:

1. UAV communication architecture and protocols based on IoT;
2. 5G-enabled UAV communications;
3. Best UAV deployment strategies;
4. UAV control techniques and methods with low latency;
5. High-precision navigation techniques;
6. Techniques for real-time surveillance;
7. Analysis of UAV network capacity;
8. UAV network ad hoc networking, routing, handover, and meshing;
9. Multiple UAVs working together to achieve a common goal; and
10. UAV for real-time applications.

AI has become increasingly important in many real-world applications. Furthermore, the need for AI has caused a lag in R&D activities. In particular, the limitation on applying for UAV has been due to multidisciplinary work. Drones are used for various purposes worldwide, including transporting medical facilities and mapping agricultural fields and wildlife. Drones were unique not only for their ability to fly at low altitudes. Drones have become as valuable as dynamic sensors with no human interaction due to their ability to capture data from remote locations, monitor videos, and give various solutions. These autonomous flying objects can understand and act in their surroundings by integrating AI and deep learning. The system's advanced algorithms can detect search objects instantly by scanning data from multiple cameras and using computer vision. Another application of AI drones with predictive capabilities is the use of historical data.

Precision agriculture, infrastructure, and construction projects can benefit from unmanned aerial vehicles (UAVs) and machine learning in data mining. For example, models used in the rain-runoff process include convolutional Artificial Neural Networks (ANN) models, which predict rainfall, estimate evaporation, predict plant water uptake, identify plants based on leaves, and assess crop production. In addition, the visual data from the UAV in the biological process can be evaluated. Figure 3 depicts the revenue generated by drones from 2016 to 2025. The most difficult challenge for the commercial drone industry is air traffic management in shared airspace. With an exponentially increasing number of drone fights, conventional air traffic management has proven ineffective in managing aerial traffic. While these

Figure 3. Revenue of drones during the period of 2016 to 2025

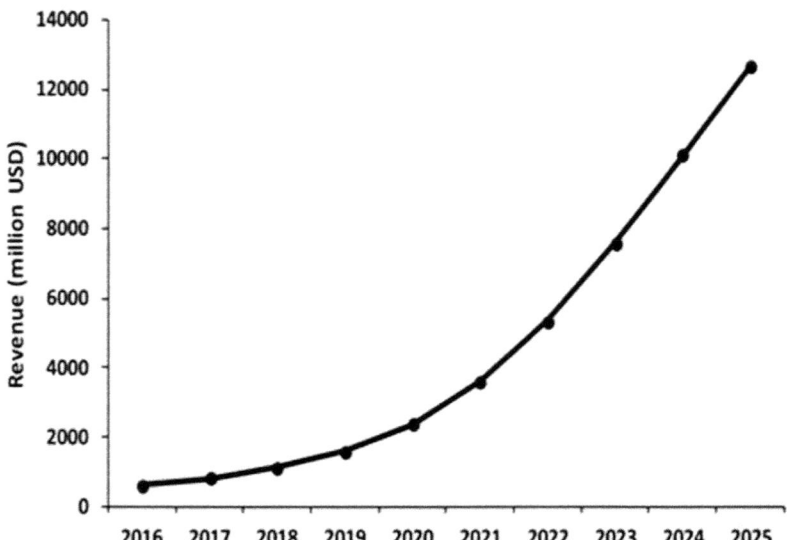

methods worked well for flights, more is needed for autonomous drone flights (Suzuki et al., 2016).

Digital twins in engineering, health, and urban planning, but in most cases, each twin is a one-of-a-kind, proprietary solution that only works with a single application. The digital twin detects changes in the general state of the UAV over time (due to mechanical wear and tear and logged flight time, among others) and updates its state to match the physical UAV. This updated digital twin can then predict how the UAV will evolve, allowing the physical asset to be directed most effectively in the future. The graphical model enables each digital twin to be "built on the same underlying computational model, but each physical asset must retain a distinct 'digital state,' which creates a unique model configuration." UAV for everything from calibration studies to a simulated "minor damage" scenario. Its digital twin extracts information from sensor data, predicts how the UAV's structural health will change, and recommends maneuvering changes to accommodate those changes (Zaman et al., 2017).

Augmented reality (AR) and virtual reality (VR) in unmanned aerial vehicles (UAVs) are popular technologies that can improve UAV operation by providing new ways to give a digital context and feedback over a depiction of a real scenario. This section discusses constructing AR experiences, the technology required, and Web AR as a platform-agnostic alternative. It also highlights the current state of augmented reality and provides examples of its application. Expand augmented reality (AR) and virtual reality (VR) in the drones, robots, entertainment, tourism, and gaming industries. Users of virtual reality equipment are immersed in a simulated environment via sight and sound, giving the impression that they are in another place entirely. AR augments the real world with intelligent data by digitally improving existing environments, such as 3D maps or scores and plays overlaid on sports telecasts, to make them more interactive and meaningful to the user. AR and VR devices are expected to overgrow, with growth rates comparable to smartphones. Devices must become smaller, more compact, and more fashionable to achieve this level of growth. Simultaneous location and mapping (SLAM) is a must-have feature for AR/VR applications. This entails room mapping and extracting features from the scene around the user for positional tracking. Drone Collision Avoidance As collision-avoidance technology has reduced crashes, consumer drones for aerial photography have grown in popularity.

Image sensors with ultra-high-resolution 4K video and global shutters are well-suited to drone requirements. Pass-Through Viewing for VR Headsets Pass-through viewing for VR headsets allows users to move around the real world while playing in the virtual world without colliding with objects. Two imaging methods enable fast frame rate and global shutter technology with RGB sensors. Human-Computer Interaction Gesture-tracking capabilities in VR/IR applications enable human-computer interactions. Infrared (IR) light sources illuminate the area near the head-mounted display, while global shutter sensors with high near-IR (NIR) sensitivity capture hand motions.

CONCLUSION

In this chapter, we present a conceptual model of how such an architecture can be developed, as well as the components of an Internet of Drones system based on Artificial Intelligence, Machine Learning, and Digital twin are discussed. The future of drones will focus on the thrust area of computer-based domains. The way we live has changed dramatically due to the use of the web, i.e., vivified cooperation among people at a computer-generated standard in contexts ranging from professional lifespan to public interactions. The IoT enables connections between making things, allowing exchanges to occur whenever, wherever, and among anything. To accomplish this, demonstrate the IoT as an unbounded web component. Depending on the needs of prospective clients, IoT applications offer a variety of benefits, including the implementation of smart cities; smart networks; intelligent transportation; innovative industrial facilities and assembly; healthcare; learning; and agriculture. Furthermore, IoT has been used to address UAV challenges such as cost reduction, improved administration, and superior security. This chapter has also presented a system for making productive choices because making the right decisions is critical, and the consequences can range from innocuous to deadly. Later, we will propose new models for deploying IoT in various critical industries, including medical services, transporting goods during emergencies, and gardening or agricultural pesticide spraying.

ACKNOWLEDGMENT

The authorship team acknowledges all the staff members of Aerospace Engineering, Chandigarh University, India, for their constant support and encouragement to conduct the research. However, this research received no specific grant from any funding agency in the public, commercial, or not-for-profit sectors.

REFERENCES

Alipour-Fanid, A., Dabaghchian, M., Wang, N., Wang, P., Zhao, L., & Zeng, K. (2019). *Machine Learning-Based Delay-Aware UAV Detection and Operation Mode Identification over Encrypted Wi-Fi Traffic*. arXiv preprint arXiv:1905.06396.

Chen, M., Saad, W., & Yin, C. (2019). Liquid state machine learning for resource and cache management in LTE-U unmanned aerial vehicle (UAV) networks. *IEEE Transactions on Wireless Communications*, *18*(3), 1504–1517. doi:10.1109/TWC.2019.2891629

EzumaM.ErdenF.AnjinappaC. K.OzdemirO.GuvencI. (2019). Micro-UAV Detection and Classification from RF Fingerprints Using Machine Learning Techniques. doi:10.1109/AERO.2019.8741970

Ge, X., Wang, J., Ding, J., Cao, X., Zhang, Z., Liu, J., & Li, X. (2019). Combining UAV-based hyperspectral imagery and machine learning algorithms for soil moisture content monitoring. *PeerJ*, *7*, e6926. doi:10.7717/peerj.6926 PMID:31110930

Jónsson, S. (2019). *RGB and Multispectral UAV image classification of agricultural fields using a machine learning algorithm*. Student thesis series INES.

Khairunniza-Bejo, S., Mustaffha, S., & Wan, I. W. I. (2014). Application of artificial neural network in predicting crop yield: A review. *Journal of Food Science and Engineering*, *4*(1), 1.

Khosravi, V., Ardejani, F. D., Yousefi, S., & Aryafar, A. (2018). Monitoring soil lead and zinc contents via combination of spectroscopy with extreme learning machine and other data mining methods. *Geoderma*, *318*, 29–41. doi:10.1016/j.geoderma.2017.12.025

Kwak, G.-H., & Park, N.-W. (2019). Impact of Texture Information on Crop Classification with Machine Learning and UAV Images. *Applied Sciences (Basel, Switzerland)*, *9*(4), 643. doi:10.3390/app9040643

Liu, X., Liu, Y., Chen, Y., & Hanzo, L. (2018). *Trajectory design and power control for multi-UAV assisted wireless networks: a machine learning approach*. arXiv preprint arXiv:1812.07665.

Mandloi & Inada. (2019). Machine Learning Approach for Drone Perception and Control. In *International Conference on Engineering Applications of Neural Networks*. Springer.

Naveen Kumar, K., Prasad, G., Rajasekar, K., & Vadivelu, P. (2018, October). Satyanarayana Gollakota. Kavinprabhu S. K, A Study On The Forest Fire Detection Using Unmanned Aerial Vehicles. *International Journal of Mechanical and Production Engineering Research and Development*, *8*(7), 165–171.

Peng, J., Biswas, A., Jiang, Q., Zhao, R., Hu, J., Hu, B., & Shi, Z. (2019). Estimating soil salinity from remote sensing and terrain data in southern Xinjiang Province, China. *Geoderma*, *337*, 1309–1319. doi:10.1016/j.geoderma.2018.08.006

Prasad, G. (2020). Performance estimation of twin propeller in unmanned aerial vehicle. *INCAS Bulletin, 12*(2), 143–149. . doi:10.13111/2066-8201.2020.12.2.12

Prasad, G., Abhishek, P., & Karthick, R. (2018). Influence of Unmanned Aerial Vehicle in Medical Product Transport. *International Journal of Intelligent Unmanned Systems, 7*(2), 88-94. . doi:10.1108/IJIUS-05-2018-0015

Prasad, G., Vijayaganth, V., Sivaraj, G., Rajasekar, K., Ramesh, M., Raj, R. G., & Matheeswaran, P. (2018). Positioning of UAV using Algorithm for monitoring the forest region. *IEEE Xplore Digital Library*, 1361–1363. doi:10.1109/ICISC.2018.8399030

Raja Sekar, K., Vignesh Moorthy, D., Prasad, G., Manigandan, P., & Senthil Kumar, M. (2018, October). An Experimental Investigation On The Influence Of Bio-Inspired Tubercle In The Unmanned Aerial Vehicle Propeller. *International Journal of Mechanical and Production Engineering Research and Development*, *8*(7), 404–409.

Rey, N. (2016). *Combining UAV-imagery and machine learning for wildlife conservation.* Academic Press.

Suzuki, T., Tsuchiya, T., Suzuki, S., & Yamamba, A. (2016). Vegetation Classification Using a Small UAV Based on Superpixel Segmentation and Machine Learning. *Journal of The Remote Sensing Society of Japan, 36*(2), 59–71.

Wagner, I. (2019). Projected commercial drone revenue worldwide 2016-2025. Statista GmbH.

Zaman, B., Mac Mckee, D., & Jensen, A. (2017). *UAV, Machine Learning.* And GIS for Wetland Mitigation in Southwestern Utah.

Chapter 11
Carpooling Solutions Using Machine Learning Tools

Sabyasachi Pramanik
Haldia Institute of Technology, India

ABSTRACT

Many people in the world face travel issues in their daily lives, such as traffic congestion, lack of parking spaces, fuel, waste, and pollution. For example, if there are not enough parking spaces on a university campus, it will take a significant amount of time for the students as well as faculty to find parking. In addition, if parking is available outside, employees will have to travel a long distance to enter the building, which will require additional time. Each day, many people in the world must travel a considerable distance to get to their destination. The purpose of this chapter is to resolve the travel issue by proposing a carpool solution using machine learning techniques.

INTRODUCTION

India boasts the world's second-largest geographical area and the world's second-largest road network. By 2020, the country's road network will have covered about 5.89 million kilometers. Moreover, 90 per cent of India's total passenger traffic of India utilizes the road network for communication (Samanta et al., 2021; Anand et al., 2022) and transportation. At the same time, the world's vehicle population is more than 0.66 kilometers per square kilometer.

Carpooling (Ma et al., 2021) is when a group of people wish to use their automobile to go from one home to another, not only to save money but also to preserve the environment from pollution by using less gasoline and also to provide them joy

DOI: 10.4018/978-1-7998-8805-5.ch011

while traveling. The carpooling services supplied via the website will operate as a barrier between various unknown persons who wish to travel in a shared manner and will only need calculating tips when collective displacements from their source to each other's destination are required. Carpooling is a socially acceptable and environmentally beneficial method of sharing travel. It aids in reducing carbon dioxide (co2) emissions, reducing traffic congestion, and resolving different parking space concerns, allowing us to promote carpooling during times of rising fuel costs and high pollution levels. Because of the fast expansion in transportation, providing common transportation services has become more difficult. Carpooling also helps to save transportation costs such as gasoline, maintenance, toll fees, and the hardship of driving by allowing ordinary people to use one vehicle for communication.

LITERATURE REVIEW

The Internet of Things (IoT) uses various devices capable of talking and exchanging information since this concept allows them to be used in both active and passive configurations. The essential goal of IoT (Liu et al., 2021) is to create clever, intelligent settings or places where objects such as smart cities (Kaushik et al., 2021), smart homes, and smart transportation (Stiles, J. et al., 2021) are self-aware for unique and inventive applications. On the IoT, each item and entity (thing) is given a unique identification to receive or send data mechanically from or to a network. Radio-Frequency Identification (RFID (Shaohao, X., et.1al. 2020) sensors (Wei et al., 2021), actuators (Zenkour et al., 2021), detectors (Gonzo, 2021), and other IoT gadgets are examples. Much of the IoT consists of various intelligent computer devices and associated sensor systems that are primarily utilized in vehicle-to-vehicle (V2V) and machine-to-machine (M2M) communication, as well as wearable computing devices for various reasons. The IoT is steadily expanding its application sector in several technology domains, securing its position in transportation and traffic management. The main issue in the present situation is the rise in the number of passenger vehicles (i.e. automobiles), which is directly proportional to population growth.

Consequently, substantial issues such as extreme traffic congestion (Moyano et al., 2021), accidents, noise, and travel time lag have arisen. Car sharing is an improved means of transportation that allows numerous people to share a single car trip, regardless of their origin. The main goal is to minimize the overall number of vehicles on the road at any time while lowering the travel cost for each rider. Car sharing is, in reality, the most frequent general para transit method, in which passengers from various user groups share a vehicle that follows their pre-determined itinerary. In addition, car sharing facilitates ridesharing using passenger automobiles, based on sharing a single-passenger vehicle with others, with the vehicle's owner being a third party (subject to business).

Consequently, most urban users or passengers can travel in a shared vehicle without owning one. They may do so after acquiring a standardized key card from one of the expressly named approved stations and paying a predetermined wage or reimbursement cost. The IoT provides a horizon for automobile sharing compared to the existing scenario. The main objective of the IoT is to connect everyone to everything, 24 hours a day, seven days a week, and from any location, which is critical for the car-sharing system.

CHAPTER FOCUS

Congestion caused by automotive traffic is one of the primary challenges in our daily lives in most cities and towns. As a result, quality of life suffers, with negative economic consequences (Baker et al., 2021), social welfare, and the environment (Pradhan et al., 2022). As a result, it will need a lot of work and dedication to study and establish imaginative and determined transportation modes in metropolitan areas to achieve a car-free lifestyle, which is one of the primary reasons for city traffic congestion. It is now difficult to imagine a sustainable transportation strategy to meet our demands. As a result, major cities throughout the globe are confronted with issues such as reducing greenhouse gas emissions and air pollution caused by car traffic congestion. As a result, combining several modes of transportation (also known as combination transport or multimodal transport) is still a big demand in metropolitan areas. To assess the social, environmental, and climate change implications, we must examine the vehicles utilized, the vehicle's energy sources, and the infrastructure necessary for transportation implementation. Public transportation infrastructure is one of the greatest ways to deal with and control traffic congestion caused by vehicles. However, there are several drawbacks, such as: replacing software, hardware, or mechanical components in a city's public transportation system is exceedingly expensive.

Furthermore, public transportation usually follows a defined geographic itinerary and schedule. As a result, there needs to be more room for change in a particular area of the public transportation system. As a result, many individuals choose to drive their vehicles or use other private transportation methods over public transit.

In recent years, the outcomes of carpooling have gained enormous notoriety and recognition in the field of sustainable transportation. Carpooling is the sharing of automobile travels in which numerous passengers may travel in a personal vehicle along the same route. As a result, it combines the convenience of owning a vehicle with the cost-effectiveness of public transit. Two ways are often employed in a vehicle trip to utilize it for sharing reasons, among other things:

1. In the first method, many passengers who wish to travel the same route may share a seat in their automobile by swapping drivers; and
2. When a driver wants to assign an unused seat in their vehicle among potential passengers, the seats are merged for payment towards fuel expenditures, with the driver's expenses being returned at the end.

In contrast to the above two techniques, the first one is widely utilized among users who wish to travel on the same route with the same source and destination and on the same timetable. In the following instance, all participants continued to make arrangements regarding the shared ride's details. However, the second technique informs the traveler of a route suited for the infrequent traveler.

The utilization of carpooling services has a wide variety of beneficial effects, including:

1. It essentially means lowering the cost of travel for the driver;
2. For the rider, it essentially means more adaptability at a lower cost, making it an excellent alternative for public transit;
3. In terms of the environment, it is primarily intended to reduce air pollution by reducing the number of vehicles used in transportation;
4. For cities, it is primarily intended to alleviate traffic congestion and the requirement for car parking space; and
5. The shared cost of the journey should be accurately apportioned between the participating driver (who shares their automobile) and the passengers, lowering the overall trip cost for each of them. It also implies that the driver would be able to earn money and recoup his expenditures.

Carpooling services have acted as a bootstrap for our transportation system in recent years, and this has only been possible thanks to the inclusion of web-based platforms that allow passengers and drivers to interact and make necessary arrangements for a shared ride, including the arrangement of a trip schedule with their path on or before the trip. Carpooling services are now accessible as smartphone and tablet apps for hand-held contemporary mobile devices, which are becoming more popular in our culture. Because people are used to accessing the Internet using these devices, there are no concerns with using them to access carpooling services. However, the availability of these devices is solely dependent on the availability and utilization of a high-quality remote sensor that can deliver context-based information at any time and in any location. A smartphone-based carpooling application requires a mobile network method to connect and communicate with passengers and drivers interested in traveling along comparable routes. The essential principle of this network is that each ride request or offer information must reach many network members planning to travel, i.e. the chances of finding a travel companion should be high. As a result, special appreciation is extended to the network's information supplier. However, users may also deny the information if they do not want to share their automobile with individuals uninterested in their other circumstances. This sort of problematic scenario in carpooling services causes roadblocks to the global expansion of these businesses (Pramanik et al., 2020). As a result, most current carpooling systems do not perform as well as expected and must be improved.

KEY ASPECTS OF CARPOOLING

In this case, comparing and matching ride requests with offers are only based on temporal and spatial parameters. Therefore, knowing the specifics of categorizing the temporal and positional aspects of carpooling is critical. In this case, the categorization may be summarized as follows:

1. The term "identical ridesharing" refers to when the rider's source pickup and destination drop-off locations are identical to the driver's source and destination locations;
2. When source pickup and destination drop-off locations are on the driver's route, inclusive ridesharing is expected to occur. However, shared trips cover only a portion of the driver's journey;
3. When the source pick-up and destination drop-off locations are on the driver's original route but do not correspond with the rider's source and destination locations, it is said to be partial ridesharing; and
4. Detour ridesharing is defined as when a driver must divert their route to the rider's source location for pickup.

The departure time specified by the drivers and the passengers, which most of the time fluctuates, is the real temporal factor that causes irregularities in the whole carpooling process. It is crucial to match users based on this attribute as it is critical to determine the actual rider before a ride's commencement. Here, a match is created between the rider's requests and the offers offered by the car-pooler service provider, which is only based on the geographical distance between the rider's route, determining the pick-up and drop-off times. Another important temporary feature is the flexibility in pick-up timings, which indicates how long the riders are willing to wait at the time of pick-up and which should be acceptable to everybody without interfering with the rider's trip. Planning and discovering pick-up and drop-off times for similar ridesharing is incredibly simple with little effort and time. However, for other ridesharing services, determining the meeting times via which the car-pooler chooses the estimated arrival time at the pick-up place takes a lot of work. Another important temporary factor to consider when calculating the rider-driver matching time is the previous calculation of journey length, which should be determined only on a temporal basis.

Solutions for Stationary Conditions

To use carpooling applications, the intended user must create a new account by filling out the required information and then providing their ride requirements, such as: desired source; destination pick-up; drop-off locations; desired pickup time; and optionally required time flexibility as per their needs during their trip. Carpooling apps are often run on third-party social networks or via their websites or apps to boost user trust and trustworthiness and enable people to participate in social activities with their senses. In addition, a rating system is employed in certain carpooling apps, allowing users to comment on each other after each journey to improve carpooling services more positively and healthily for all carpoolers. When carpooling, the trip expenditures may be paid in cash or by automatic payment over the Internet once they have shared the costs. Using this approach, riders may compensate drivers before or after their journey, depending on the distance they traveled in a shared vehicle.

Carpooling Apps That Are Not Mobile

The most frequently recognized ridesharing execution in the current environment is based on a "static" approach. Car-poolers submit their requests and offers in this network, which is valid for many hours for future transportation needs, and all criteria should be addressed before the start of journeys. In most cases, while approaching static carpooling, the different possibilities of unanticipated changes in date, time, or other aspects of the shared journey are not considered. The carpooling system in this approach is typically based on the matching of lists that are generally based on the common source and destination matching, where the riders and drivers are expected to communicate regularly to make proper arrangements for upcoming trips and then reach an agreement for the same. The agreement between them should contain the rider's pickup and drop-off locations specifications, their required meeting time, the actual price incurred for riding, and the possibility of carrying baggage.

In most cases, the static carpooling technique produces better results for long-distance trips planned ahead of time. However, it could perform better in metropolitan locations where consumers want additional services and amenities. BlaBlaCar

(Wagih et al., 2021) is now Europe's static carpooling service's most frequent and well-known case study.

BLABLACAR: A CASE STUDY

This section provides comprehensive illumination for a static carpooling method by detailing the procedure in depth using BlaBlaCar. The software provider offers this static ridesharing program for resident users through internet-enabled mobile phones and desktop web browsers. While writing this chapter, static carpooling process was the world's largest long-distance ridesharing community, with over 60 million members from 22 countries across four continents. BlaBlaCar connects drivers who want to drive and passengers who really want to travel with them. These customers usually wish to organize shared transportation between two cities or within a city and split the travel expense. The targeted users must first register on the desired website and then establish their profile, including information such as the user's hobbies, profile responsibilities, a community network profile, and rating and reviewing. All of its services are available to customers through mobile phones, websites, and applications for Android and iOS platforms. BlaBlaCar, like other static ridesharing platforms, needs an agreement for scheduled long-distance travel, where drivers desire to fill empty seats with passengers for extended excursions. Offers to the rider in this service may be made just as they can to the driver or passenger. The program lets users specify source and destination locations by entering or choosing from a list of pre-defined locales. It necessitates the usage of a GPS-enabled mobile phone, via which the user may establish their current position before the commencement of the voyage and optionally add one or more waypoints. The driver and rider's route information is then matched to see who has a similar source, destination, or waypoint. These programs enable users to set the desired departure time and the required payback amount at the source location. The user may then choose his travel companion from a matching list based on their choices, such as source, destination, and timings, as well as the driver's payback amount. When everyone agrees to a shared trip, the driver will pick up the passenger at the agreed-upon time and location. The overall cost of the voyage is shared among the participants, who are responsible for making their payment by phone or PayPal.

One of the key drawbacks of this strategy is that it does not allow for temporary trip management, and the characteristics are as follows:

1. Initially, users were compared based on source and destination pairs and waypoints. As a result, the same ridesharing mode is available here;

2.   If feasible, shared transport may be scheduled in advance, from a few hours to a few days; and

3.   During the journey, the driver cannot pick up any riders that cross their route. Therefore, it is only conceivable if their journeys are perfectly matched if they have similar stopovers.

Ridesharing in a Dynamic Mode

Dynamic ridesharing is regarded as a new sort of carpooling than static carpooling since it is more desirable in metropolitan areas. In this ridesharing system, the provider used an automated approach to make matches between drivers and passengers in a short time. The timing begins here within a very short time gap, ranging from a few minutes to many hours before the actual departure time from the source side. It is the most recent and advanced sort of ridesharing technique, which is being used by many individuals. It outperforms the standard static ridesharing model in several ways.

1.   Information and communication technology are crucial aspects that play a significant and sensitive role in allowing dynamic ridesharing from a technological standpoint.

2.   Ridesharing apps that are always changing.

When we think about its implementation, we need much new technology to exploit the numerous services given by dynamic ridesharing. The following are some of the most widely utilized technologies:

a.   Both passengers and drivers utilize mobile devices (smartphones) to interact with and access numerous services, as well as to organize everything for a shared trip when it is in portable mode;

b.   A GPS gadget (often found within a smartphone) used to monitor the driver's movement to determine the whereabouts of passengers; and

c.   The usage of social media is being used to raise the level of honesty between the driver and the passenger.

Constant network connection necessitates that all shared ride arrangements be completed promptly, which may require the user's mobile device to be connected to the Internet 24 hours a day, seven days a week. Many global telecommunication firms now offer such services with a continuous mobile connection without fail. The aforementioned current technologies are crucial enablers for the smooth operation of dynamic ridesharing. Though ridesharing projects have been launched in the past, the primary challenge at the time was the need for such communication tools. All the elements mentioned above are managed by a reliable network service that can instantly connect a rider with a driver to begin a journey and assist with the payment process using suitable optimization algorithms (Pramanik et al., 2020; Pramanik et al., 2014). As a result, the following key characteristics of Dynamic Ridesharing have been identified:

1. All of the arrangements for shared transport may be made on short notice, making it dynamic;
2. Independent drivers who are not affiliated with any organization may participate in this dynamic ridesharing;
3. The motto of dynamic ridesharing is cost-sharing, which results in cost reductions via cost-sharing among both drivers and passengers;
4. At the end of the day, the fundamental goal of Dynamic Ridesharing is non-recurring journeys. Traditionally, carpooling necessitates starting the shared rides to work many hours in advance, and the nature of the journeys is essentially repeating, i.e. repeated trips (such as daily travel to the workplace with the help of colleagues). It is better suited for on-demand, continuous, planned share rides that may be scheduled in advance in a matter of minutes or hours;
5. The pre-planned shared trip is organized between drivers and passengers entirely in advance, without the need to go to a specific destination; and
6. Automated matching automates the whole process without requiring much additional labor. Moreover, it does so in a relatively short period with little work on the part of the participants. Finding a good partner for the journey is not done manually here; the system automatically matches the riders and drivers, allowing them to communicate.

Lyft as an Example

Lyft began in the USA as a dynamic ridesharing organization system based in San Francisco, founded by Logan Green and John Zimmer in 2012. The on-demand ridesharing network system was launched to make ride-ride services more accessible for shorter journeys within or between cities. Lyft is now utilized in many cities, such as San Francisco, Texas, Los Angeles, New Jersey, and New York City. Users must download the Lyft program from the store for their Android and iOS devices to utilize it and its services. After that, customers may join up and provide their cell phone numbers for communication and choose the best payment option for each journey. If a user wishes to use Lyft for a journey, the driver must search for a nearby location using an interactive map. When the shared ride arrangement is complete, the passenger may speak with the chosen driver to confirm the accuracy of their information, such as: their name; rating; profile photo; and vehicle type. When the journey begins, the riders are picked up from their current position by the designated driver. The overall cost of the shared transport is divided among the tour participants, who then pay using their smartphones and PayPal. The Lyft service's ideology aims to instill trust in its consumers. Riders are obliged to offer their evaluations to the driver when the journey is over, and only the highest-rated drivers are allowed to spare their seats for further trips. All participants in this procedure must undergo a screening process that includes checking their criminal history, car standards, and drug and alcohol addiction, among other things. One of the system's potential downsides is that riders are picked up from their current position by the drivers. As a result, the proper matching algorithm cannot use partial ridesharing mode. Because drivers may only pick up the passengers closest to them or within a specific distance, the driver's actual path is not considered for assessment or matching.

CARPOOLING'S SOCIAL BENEFITS

The current carpooling system offers various socioeconomic advantages, including:

1. By sharing, the number of miles driven by the vehicle is reduced;
2. Fuel consumption is reduced, as is the quantity of greenhouse gas (GHG) emissions released into the environment;
3. Reduced detrimental effects of air pollution on various societal population categories such as low-income, low-cost, and other verified environmental populations; and
4. Employers and government organizations both need to save costs.

Vehicle Miles Travelled (VMT)

It is a travel measure that indicates the total number of kilometers traveled by each vehicle in carpooling. According to research conducted by the Federal Highway Administration (FHA) in the 1970s, the energy crisis was alleviated by 23 per cent in terms of Vehicle Miles Travelled (VMT). Employee Based Trip Reduction and Transportation Demand Management programs, capable of performance monitoring and evaluation, are the best practices for supporting VMT reduction targets. Only a few hand-count empirical research may be used to investigate the effect of VMT regulations. According to one study, workers who engaged in the program had lower VMT rates, ranging from 4.2 per cent to 4.8 per cent, than employees who worked at the same location and were not required to participate in the program. When Washington State's Commute Trip Reduction Law was introduced and research was conducted on it, it was discovered that it had comparable impacts on VMT. Due to this rule, it has also been found that the average VMT decrease per employee at work locations is 6 per cent. According to some estimates, such initiatives may reduce VMT by 4 to 6 per cent for office driving (approx. 1 per cent regionally). VMT reduction estimates for a whole region or a metropolitan area were based on precisely two research studies. Essential data were obtained directly from businesses for the research to calculate the list of commute trips eliminated as per Washington State's Commute Trip Reduction (CTR) program. According to the research, overall VMT in the four core counties of metropolitan Seattle has decreased by 1.33 per cent for all routes and 1.07 per cent for all freeways. Because the research author considered all sorts of journeys during the morning peak hour, including commute travel to non-participating sites and non-commute visits, it indicates a lower effect than prior studies. The Commute Trip Reduction Task Force conducted an individual analytical study for multiple years in 2005 and predicted a 1.6 per cent reduction in overall VMT based on the same data set. According to the initial findings, people exhibit interest in carpooling due to decreased journey time and expense. This is a significant aspect that has a greater influence on the net VMT of this sort of communication mechanism.

Fuel Consumption Is Reduced

In a year, the average fuel consumption of passenger automobiles and sports-related vehicles is almost equivalent to 550 and 915 gallons, respectively. Research to improve carpooling rules shows that the most efficient technique is minimizing energy use rather than fully restricting driving. Further research was conducted in the USA to determine the possibility of reducing annual gasoline consumption from 0.80 per cent to 0.82 per cent billion gallons by adding one more passenger to every 100 automobiles. According to recent research, adding one additional passenger to every ten automobiles might result in annual fuel savings of 7.54 to 7.74 billion gallons. According to other research, carpooling may save 33 million gallons of gasoline daily if a third passenger is added to an average communicative vehicle. Carpooling has a notable regional impact on fuel savings, which may be seen. The research was conducted in the San Francisco Bay Area, where gasoline consumption was reduced from 4, 50,000 to 9 00,000 gallons per year. It is related to a drop in regular traffic due to the reduction in carpooling congestion.

Emissions of Greenhouse Gases (GHGs) Are Reduced

Many researchers have intended to reduce greenhouse gas emissions in carpooling systems by reducing the amount of gasoline used. For example, when a rider joins the employer trip reduction program, a simulation model is provided for predicting each car-pooler rider to cut personal travel GHG emissions from four to five per cent. According to research conducted in the USA, adding one passenger to every 100 vehicles would result in an annual reduction of 7.2 million tonnes of GHG emissions. According to this analysis, adding one passenger to every ten vehicles would reduce 68 million tonnes of GHG emissions annually. Another research, according to the SMART 2020 report, estimates that if we employ information and communication technologies (ICT) for carpooling, such as app-based carpooling, co2 emissions may be reduced by 70 to 190 million tonnes per year.

Savings for Government Agencies and Employers

The most cost-effective strategy for carpooling is to improve infrastructure capacity and person throughput to alleviate traffic congestion and reduce the need for additional roadway and public transportation capacity. In the City of Seattle, a commute trip reduction ordinance was passed in 2017, contributing to an 11 per cent reduction in single-occupant vehicle trips. Another study was conducted on casual carpooling. The potential capacity for such carpooling was identified as reducing energy consumption for approximately 150 commuters by providing them with a standard express bus service while lowering costs.

Financial and Tax Benefits

Carpooling provides a variety of tax and financial advantages to those who participate in it, including businesses and workers. According to the 132(f) provision of the Internal Revenue Code of the USA, companies may give their employees: tax-free parking; vanpool service; public transportation; and cycling expenditures. The monthly parking loT, vanpool service, and public transportation benefit maximum were US$260 per month, subject to an annual cost of living increase. Previously, companies paid such perks, but now such commuting expenditures might be removed from the employee's subsidy part by the company. According to the U.S. Levy Cuts and Jobs Act of 2017, the government's tax on such activities was repealed. However, the employer may still support these costs since there are no means to withhold the subsidized share of the commuter's costs. Employees ultimately gain from tax and financial incentives to encourage carpooling.

TRENDS AND INNOVATIONS

People's lifestyles are changing fast these days. Thus, a variety of variables influence how they travel and carpool. However, some critical elements are linked to specific sectors, such as technical, mobility, social, and demographic changes.

1. Technological developments.
2. Due to the introduction of many new tools and technologies, i.e., technical trends, carpooling has become increasingly popular in the contemporary day.
3. Cloud computing, geographical area-based navigation systems with associated services, and current mobile phone-based communication technology with associated computer capabilities are some of the most recent technology trends.
4. Growth in data availability, compilation, dissemination, accumulation, and re-broadcasting by cloud source public and private sectors is enabled through public-private partnerships, APIs, and additional tools.
5. Carpool passenger services are more convenient when they enable app-based and on-demand transportation choices.

Trends in Mobility

Due to a rise in travel demand, well-managed urban congestion, a decrease in financing for transportation facilities, and the urgent need to expand current infrastructure facilities, the demand for dispersed and improved occupancy options such as micro-transit, app-based carpooling and many more is increasing. Due to customer demand, the number of on-demand transportation options has increased.

Social Developments

There is an increase in environmental awareness around carbon emissions. Consider a huge region as an economic hub and draw transportation lines around it. As one goes toward urbanization, one will rely on private automobiles. With the aid of mobile Internet, there is a growing demand to receive rapid results and services.

Trends in Demography

People wish to labor for extended periods without regard for the distance traveled due to population growth and changing lifestyles. When a person's impairment in society grows, they will desire to work to support themselves. As the number of the family grows, they will wish to communicate via their private transportation cars.

People in society willing to work after they retire from their jobs not just to keep themselves engaged (keeping fit) but also to generate money.

BENEFITS OF CARPOOLING

1. This reduces travel costs and the likelihood of obtaining a personal vehicle. In addition, a lone traveler may receive different companies, and the fuel price can be divided by taking a shared journey.
2. It cuts down on journey time. Because shared journeys are quicker on highways, and if there is more than one passenger in the vehicle, the automobile will only go in the lanes with the least traffic. It not only allows one to choose the best route for one's needs, but it also allows one to travel on a budget.

Strategies for Carpooling

Carpooling, often known as car-sharing, is a cost-effective way to save money while simultaneously lowering pollution. It is well-known and widely used in advanced European nations such as the USA, France, Germany, Italy, and Spain. In India, they put in place various methods to encourage carpooling. To embrace carpooling, the following methods were necessary.

1. It is necessary to legislate for carpooling to ensure the legal status of the process and protect the interests of car-poolers and their legal rights.
2. By creating a specific carpooling organization to lead carpooling disinformation, association, and service activities, the government may provide the groundwork for the process. It also encourages public carpooling institutions to aid in the growth of carpooling.
3. Initiate and execute a carpooling incentive campaign to increase the percentage of people that carpool daily.
4. Begin a carpooling pilot study to determine the program's effects and utility.
5. Carpooling process management depends on well-managed operations, such as determining an accurate pickup time for a specific route after considering all factors.

Carpooling Power Tools

Scala, the programming language used to construct the most popular carpooling system, is regarded as the most crucial tool. The same tools, namely Akka, are utilized to create concurrent applications. Another toolkit, OscarR, is used to handle Operations Research issues, and its most notable features are a constraint programming solver and visualization capabilities. Finally, the travel time between two places was calculated using Google Maps and the MapQuest API.

Scala

The term Scala (Norwicki et al., 2021) is derived from the term scalable. This is a general-purpose, high-level programming language that was created on the apex of the Java virtual machine platform to support object-oriented programming. It includes functional programming elements and is fully compatible with Java. The most common Java features and a variety of features from other languages are included. Scala also has several extra features, including:

1. The lambda function is a type of function;
2. Types of auto-detection;
3. Making lists;
4. A calculation that takes time to be done;
5. They are matching patterns;
6. Classes regularly;
7. A system with a single root;
8. Optional arguments;
9. Arguments with names; and
10. Using Scala, it is feasible to write typical programming patterns in a short amount of time. Martin Odersky conceived and developed it in 2004.

Akka

Akka is written in the Scala programming language. Akka is a free and open-source runtime library or toolkit for JVM that provides various tools for creating powerful, reactive, concurrent, message-driven, and distributed applications. The actor is the Akka component that not only runs in distinct threads but also has the potential to execute on many machines. As a result, it employs various Actor-Based Models. For example, these may be used to get Google Maps and MapQuest online APIs (Sharon, 2021). In addition, because the requested messages are queued and correctly handled by Akka, a group of engineers may contact the distant servers simultaneously with less effort.

Operational Research in Scala (OscaR)

"Scala in OR" is what OscaR stands for. Scala toolkit or package designed specifically for JVM to solve operational research problems. We have a thriving developer community that is working on open-source projects. OscaR offers the ability to combine several packages using the following techniques:

1. Constraints Programming (CP);
2. Local Search with Restrictions (CBLS);
3. Linear Programming (LP);
4. Discrete Event Simulation (DES);
5. Optimization without Derivatives; and
6. Visualization.

For the carpooling solution, a constraint programming solver and a visualization tool: Map Quest and Google Maps are employed. The primary goal of combining Google Maps (Lavorgna, 2022) and Map Quest is to create a web API that allows a simple calculation to estimate the real-time necessary while traveling by automobile from one location to another. It may also be computed locally using the most appropriate geographical data. However, it is more readily done by utilizing various online services, requiring less time and effort. Moreover, API eliminates the hassle and stress of keeping up with the newest data on roadwork and other traffic-related information.

The following are two essential aspects that were the focus of this study for the issues mentioned above:

Time: Several modes of transportation are available (Buses, Peer-to-Peer ridesharing, private transportation). Buses and other public vehicles have set departure and return times. Each time, it navigates a large number of students and staff members. For example, if students and employees finish their job, they must return home after the bus runs according to their previous timetable. As a result, everyone must wait till the bus does not come.

Cost: Private transportation is too costly for both employees and students. Their travel costs are set within a range, i.e. within the monthly limit; nevertheless, the distance between riders fluctuates. Furthermore, there needs to be more information on the vehicle's owner. Female passengers will be unsafe in this car since it lacks security (Bandyopadhyay et al., 2021; Bansal et al., 2021; Pramanik & Raja, 2020) design. Some drivers of vehicles adjust the car seats to accommodate many students. Combining car owners with female pupils or workers has no purpose or applicability.

CONCLUSION

The main motivation for writing this chapter was experiencing daily problems, such as traffic congestion, parking shortages, fuel waste, and pollution. It would take longer to hunt for parking on a university's campus if insufficient. Employees will have to drive a considerable way to get to the building if parking is accessible outside, which will take longer. The same problem will develop for residential reasons. If a university has several research stations that are geographically separate from one another, communication occurs between them and the main office using office automobiles. However, leased vehicles from a travel agency are sometimes purchased. As a consequence, transportation costs will climb.

Similarly, all-day scholars who do not live in the dormitory and arrive from their homes are designated day scholars. As a result, students must travel a long distance in hired cars to visit the facility daily. Solutions in carpooling utilizing machine learning (Dutta et al., 2021) methods discussed in this chapter.

REFERENCES

Anand, R., Singh, J., Pandey, D., Pandey, B. K., Nassa, V. K., & Pramanik, S. (2022). Modern Technique for Interactive Communication in LEACH-Based Ad Hoc Wireless Sensor Network. In M. M. Ghonge, S. Pramanik, & A. D. Potgantwar (Eds.), *Software Defined Networking for Ad Hoc Networks*. Springer. doi:10.1007/978-3-030-91149-2_3

Baker, H. K., Kumar, S., & Pandey, N. (2021). Thirty years of *Small Business Economics*: A bibliometric overview. *Small Business Economics*, *56*(1), 487–517. doi:10.100711187-020-00342-y

Bandyopadhyay, S., Goyal, V., Dutta, S., Pramanik, S., & Sherazi, H. H. R. (2021). Unseen to Seen by Digital Steganography. In S. Pramanik, M. M. Ghonge, R. Ravi, & K. Cengiz (Eds.), *Multidisciplinary Approach to Modern Digital Steganography*. IGI Global. doi:10.4018/978-1-7998-7160-6.ch001

Bansal, R., Jenipher, B., Nisha, V., Makhan, R., Pramanik, S., Roy, S., & Gupta, A. (2022). Big Data Architecture for Network Security. In Cyber Security and Network Security. Wiley. doi:10.1002/9781119812555.ch11

Dutta, S., Pramanik, S., & Bandyopadhyay, S. K. (2021). Prediction of Weight Gain during COVID-19 for Avoiding Complication in Health. *International Journal of Medical Science and Current Research*, *4*(3), 1042–1052.

Gonzo, R., & Pokraka, A. (2021). Light-ray operators, detectors and gravitational event shapes. *J. High Energy. Phys., 15*. doi:10.1007/JHEP05(2021)015

Kaushik, D., Garg, M., Annu, Gupta, A., & Pramanik, S. (2021). Application of Machine Learning and Deep Learning in Cyber security: An Innovative Approach. InGhonge, M., Pramanik, S., Mangrulkar, R., & Le, D. N. (Eds.), *Cybersecurity and Digital Forensics: Challenges and Future Trends*. Wiley.

Lavorgna, L., Iaffaldano, P., Abbadessa, G., Lanzillo, R., Esposito, S., Ippolito, D., Sparaco, M., Cepparulo, S., Lus, G., Viterbo, R., Clerico, M., Trojsi, F., Ragonese, P., Borriello, G., Signoriello, E., Palladino, R., Moccia, M., Brigo, F., Troiano, M., ... Bonavita, S. (2022). Disability assessment using Google Maps. *Neurological Sciences, 43*(2), 1007–1014. doi:10.100710072-021-05389-7 PMID:34142263

Liu, S., Liu, X., Wang, S., & Muhammad, K. (2021). Fuzzy-aided solution for an out-of-view challenge in visual tracking under IoT-assisted complex environment. *Neural Computing & Applications, 33*(4), 1055–1065. doi:10.100700521-020-05021-3

Ma, N., Zeng, Z., Wang, Y., & Xu, J. (2021). Balanced strategy based on environment and user benefit-oriented carpooling service mode for commuting trips. *Transportation, 48*(3), 1241–1266. doi:10.100711116-020-10093-0

Nowicki, M., Górski, Ł., & Bała, P. (2021). PCJ Java library as a solution to integrate HPC, Big Data, and Artificial Intelligence workloads. *Journal of Big Data, 8*(1), 62. doi:10.118640537-021-00454-6

Pramanik, S., & Bandyopadhyay, S. K. (2014). Image Steganography Using Wavelet Transform and Genetic /Algorithm. *International Journal of Innovative Research in Advanced Engineering, 1*, 1–4.

Pramanik, S., Ghosh, R., Ghonge, M., Narayan, V., Sinha, M., Pandey, D., & Samanta, D. (2020). A Novel Approach using Steganography and Cryptography in Business Intelligence. In A. Azevedo & M. F. Santos (Eds.), *Integration Challenges for Analytics, Business Intelligence and Data Mining* (pp. 192–217). IGI Global.

Pramanik, S., & Suresh Raja, S. (2020). A Secured Image Steganography using Genetic Algorithm. *Advances in Mathematics: Scientific Journal, 9*(7), 4533–4541.

Samanta, D., Dutta, S., Galety, M. G., & Pramanik, S. (2021). A Novel Approach for Web Mining Taxonomy for High-Performance Computing. *The 4th International Conference of Computer Science and Renewable Energies (ICCSRE'2021)*. 10.1051/e3sconf/202129701073

Sharon, T. (2021). Blind-sided by privacy? Digital contact tracing, the Apple/Google API, and big tech's newfound role as global health policymakers. *Ethics and Information Technology*, *23*(S1), 45–57. doi:10.100710676-020-09547-x PMID:32837287

Sinha, M., Chacko, E., Makhija, P., & Pramanik, S. (2021). Energy Efficient Smart Cities with Green IoT. In Green Technological Innovation for Sustainable Smart Societies: Post Pandemic Era. Springer.

Wagih, H. M., & Mokhtar, H. M. O. (2021). Ridology: An Ontology Model for Exploring Human Behavior Trajectories in Ridesharing Applications. In M. Al-Emran, K. Shaalan, & A. Hassanien (Eds.), *Recent Advances in Intelligent Systems and Smart Applications. Studies in Systems, Decision and Control* (Vol. 295). Springer. doi:10.1007/978-3-030-47411-9_30

Xie, S., Zhang, F., & Cheng, R. (2021). Security Enhanced RFID Authentication Protocols for Healthcare Environment. *Wireless Personal Communications*, *117*(1), 71–86. doi:10.100711277-020-07042-6

Zenkour, A. M., & El-Shahrany, H. D. (2021). Hygrothermal forced vibration of a viscoelastic laminated plate with magnetostrictive actuators resting on viscoelastic foundations. *International Journal of Mechanics and Materials in Design*, *17*(2), 301–320. doi:10.100710999-020-09526-6

Compilation of References

Aapaoja, A., & Haapasalo, H. (2014). *The Challenges of Standardization of Products and Processes in Construction*. Research Gate.

Abdul Rahman, N., Yaacob, Z., & Mat Radzi, R. (2016). The Challenges Among Malaysian SMEs: A Theoretical Perspective. *World Journal of Social Sciences*, 6, 124–132.

Adams, R. J., & Khoo, S.-T. (1996). *QUEST: The Interactive Test Analysis System* (Vol. 1). Australian Council for Educational Research. [Software User Manual]

Agarwal, D. (2019). *Top 5 trends that will rule fashion retail in 2019*. Available at: www.indianretailer. com/article/whats-hot/trends/Top-5-trends-that-will-rule-fashion-retail-in-2019. a6270/

Agarwal, P., Vempati, S., & Borar, S. (2018). *Personalizing similar product recommendations in fashion e-commerce*. arXiv preprint arXiv:1806.11371.

Aggarwal, C. C. (2016). *Recommender Systems* (Vol. 1). Springer. doi:10.1007/978-3-319-29659-3

Ahluwalia, R., Unnava, H. R., & Burnkrant, R. E. (2001). The moderating role of commitment on the spillover effect of marketing communications. *JMR, Journal of Marketing Research*, 38(4), 45870. doi:10.1509/jmkr.38.4.458.18903

Akerman, A. (2018). A theory on the role of wholesalers in international trade based on economies of scope. *The Canadian Journal of Economics. Revue Canadienne d'Economique*, 51(1), 156–185. doi:10.1111/caje.12319

Akundi, A., Euresti, D., Luna, S., Ankobiah, W., Lopes, A., & Edinbarough, I. (2022). State of Industry 5.0—Analysis and Identification of Current Research Trends. *Applied System Innovation*, 5(1), 27. doi:10.3390/asi5010027

Alipour-Fanid, A., Dabaghchian, M., Wang, N., Wang, P., Zhao, L., & Zeng, K. (2019). *Machine Learning-Based Delay-Aware UAV Detection and Operation Mode Identification over Encrypted Wi-Fi Traffic*. arXiv preprint arXiv:1905.06396.

Al-Muharrami, S., Matthews, K., & Khabari, Y. (2006). Market structure and competitive conditions in the Arab GCC banking system. *Journal of Banking & Finance*, 30(12), 3487–3501. doi:10.1016/j.jbankfin.2006.01.006

Alsharif, M., & Nordin, R. (2017). Evolution towards fifth generation (5G) wireless networks: Current trends and challenges in the deployment of millimetre wave, massive MIMO, and small cells. *Telecommunication Systems*, *64*(4), 617–637.

Amankwah-Amoah, J., Antwi-Agyei, I., & Zhang, H. (2018). Integrating the Dark Side of Competition into Explanations of Business Failures: Evidence from a Developing Economy: Integrating the Dark Side of Competition into Explanations of Business Failures. *European Management Review*, *15*(1), 97–109. doi:10.1111/emre.12131

Ambrose, B. W., Fuerst, F., Mansley, N., & Wang, Z. (2019). Size effects and economies of scale in European real estate companies. *Global Finance Journal*, *42*, 100470. doi:10.1016/j.gfj.2019.04.004

Ameen, N. T. A., Tarhini, A., Shah, M. H., Madichie, N., Paul, J., & Choudrie, J. (2021). Keeping customers' data secure A cross-cultural study of cybersecurity compliance among the Gen-Mobile workforce. *Computers in Human Behavior*, *114*, 106531. doi:10.1016/j.chb.2020.106531

Ameyaw, B., Korang, J., Twum, E., & Asante, I. (2016). Tax Policy, SMES Compliance, Perception and Growth Relationship in Ghana: An Empirical Analysis. British Journal of Economics. *Management & Trade*, *11*(2), 1–11. doi:10.9734/BJEMT/2016/22030

Anand, D. M. (2015). *Globalization and Indian School Education: Impact and Challenges*. CORE.

Anand, R., Singh, J., Pandey, D., Pandey, B. K., Nassa, V. K., & Pramanik, S. (2022). Modern Technique for Interactive Communication in LEACH-Based Ad Hoc Wireless Sensor Network. In M. M. Ghonge, S. Pramanik, & A. D. Potgantwar (Eds.), *Software Defined Networking for Ad Hoc Networks*. Springer. doi:10.1007/978-3-030-91149-2_3

André, Q., Carmon, Z., Wertenbroch, K., Crum, A., Frank, D., Goldstein, W., Huber, J., van Boven, L., Weber, B., & Yang, H. (2018). Consumer choice and autonomy in the age of artificial intelligence and big data. *Customer Needs and Solutions*, *5*(1-2), 28–37. doi:10.100740547-017-0085-8

Asgary, N., & Walle, A. H. (2002). The cultural impact of globalization: Economic activity and social change. *Cross Cultural Management*, *9*(3), 58–75. doi:10.1108/13527600210797433

Ashraf, M., Ahmad, J., Hamyon, A. A., Sheikh, M. R., & Sharif, W. (2020). Effects of post-adoption beliefs on customers' online product recommendation continuous usage: An extended expectation-confirmation model. *Cogent Business & Management*, *7*(1), 1735693. doi:10.1080/23311975.2020.1735693

Ashraf, M., Jaafar, N. I., & Sulaiman, A. (2017). *The mediation effect of trusting beliefs on the relationship between expectation-confirmation and satisfaction with the usage of online product recommendation*. The South East Asian Journal of Management.

Atun, R., De Andrade, L. O. M., Almeida, G., Cotlear, D., Dmytraczenko, T., Frenz, P., Garcia, P., Gómez-Dantés, O., Knaul, F. M., Muntaner, C., de Paula, J. B., Rígoli, F., Serrate, P. C.-F., & Wagstaff, A. (2015). Health-system reform and universal health coverage in Latin America. *Lancet*, *385*(9974), 1230–1247. doi:10.1016/S0140-6736(14)61646-9 PMID:25458725

Auger, S. D., Jacobs, B. M., Dobson, R., Marshall, C. R., & Noyce, A. J. (2021). Big data, machine learning, and artificial intelligence: A neurologist's guide. *Practical Neurology*, *21*(1), 4–11. PMID:32994368

Ayyagari, M., Kunt, A. D., & Maksimovic, V. (2006). *How important are financing constraints?: The role of finance in the business environment*. World Bank Publications.

Aziz, N., Hossain, B., & Mowlah, I. (2018). Does the quality of political institutions affect intra-industry trade within trade blocs? The ASEAN perspective. *Applied Economics*, *50*(33), 3560–3574. doi:10.1080/00036846.2018.1430336

B20ITALY. (2021). *Finance & Infrastructure, Policy Paper 2021*. B20 ITALY. https://www.b20italy2021.org/wp-content/uploads/2021/10/B20_FinanceInfrastructure.pdf

B20italy2021. (2021). *ABOUT B20*. B20. https://www.b20italy2021.org/b20/

Baffes, J., & Gorter, H. (2022). *Experience With Decoupling Agricultur Al Support*. Social Sciences Research Network.

Baker, H. K., Kumar, S., & Pandey, N. (2021). Thirty years of *Small Business Economics*: A bibliometric overview. *Small Business Economics*, *56*(1), 487–517. doi:10.100711187-020-00342-y

Bakkar, M. (2016). *An investigation of mobile healthcare (mHealthcare) training Design for Healthcare Employees in Jordan*. Academic Press.

Bakkar, M. N., & Axmann, M. (2022). Industry 4.0: Learning Analytics Using Artificial Intelligence and Advanced Industry Applications. In *Manage Your Own Learning Analytics* (pp. 193–204). Springer. doi:10.1007/978-3-030-86316-6_9

Balabanovic, M., & Shoham, Y. (1997). Fab: Content-based, collaborative recommendation. *Communications of the ACM*, *40*(3), 66–72. doi:10.1145/245108.245124

Bandyopadhyay, S., Goyal, V., Dutta, S., Pramanik, S., & Sherazi, H. H. R. (2021). Unseen to Seen by Digital Steganography. In S. Pramanik, M. M. Ghonge, R. Ravi, & K. Cengiz (Eds.), *Multidisciplinary Approach to Modern Digital Steganography*. IGI Global. doi:10.4018/978-1-7998-7160-6.ch001

Banerji, K., & Sambharya, R. B. (1998). Effect of network organization on alliance formation: A study of the Japanese automobile ancillary industry. *Journal of International Management*, *4*(1), 41–57. doi:10.1016/S1075-4253(98)00003-9

Bansal, R., Jenipher, B., Nisha, V., Makhan, R., Pramanik, S., Roy, S., & Gupta, A. (2022). Big Data Architecture for Network Security. In Cyber Security and Network Security. Wiley. doi:10.1002/9781119812555.ch11

Bao, Y., Sun, Y., Meng, S., Shi, J., & Lu, L. (2020). 2019-Nov epidemic: Address mental health care to empower Society. *Lancet*, *395*(10224), e37–e38. doi:10.1016/S0140-6736(20)30309-3 PMID:32043982

Barber, B. R. (2013). *If mayors ruled the world: Dysfunctional nations, rising cities*. Yale University Press.

Barbic, S. B., Susan & Mayo, Nancy. (2014). Emotional vitality in caregivers: Application of Rasch Measurement Theory with secondary data to development and test a new measure. *Clinical Rehabilitation*, *29*, 29. PMID:25246610

Barua, C., Gati, B., Lajumoke, T., Taraporevala, Z., Havas, A., & Radnai, M. (2019). Beyond banking: How banks can use ecosystems to win in the SME market. McKinsey & Company, 1–13.

Basnett, Y. (2017). What do empirical studies say about economic growth and job creation in developing countries? Beck, T., & Demirguc-Kunt, A. (2006). Small and medium-sized enterprises: Access to finance as a growth constraint. *Journal of Banking & Finance*, *30*(11), 2931–2943. doi:10.1016/j.jbankfin.2006.05.009

Bathla, G., Aggarwal, H., & Rani, R. (2020). A graph-based model to improve social trust and influence for social recommendation. *The Journal of Supercomputing*, *76*(6), 4057–4075. doi:10.100711227-017-2196-2

Bellavitis, C., Filatotchev, I., Kamuriwo, D. S., & Vanacker, T. (2017). Entrepreneurial finance: New frontiers of research and practice. *Venture Capital*, *19*(1–2), 1–16. doi:10.1080/1369106 6.2016.1259733

Ben Salem, A., & Lakhal, L. (2018). Entrepreneurial coaching: How to be modeled and measured? *Journal of Management Development*, *37*(1), 88–100. doi:10.1108/JMD-12-2016-0292

Bendapudi, N., & Berry, L. L. (1997). Customers' motivations for maintaining relationships with service providers. *Journal of Retailing*, *73*(1), 15–37. doi:10.1016/S0022-4359(97)90013-0

Benlian, A., Titah, R., & Hess, T. (2012). Differential effects of provider recommendations and consumer reviews in E-commerce transactions: An experimental study. *Journal of Management Information Systems*, *29*(1), 237–272. doi:10.2753/MIS0742-1222290107

Berechman, J., & Giuliano, G. (1985). Economies of scale in bus transit: A review of concepts and evidence. *Transportation*, *12*(4), 313–332. doi:10.1007/BF00165470

Berger, N., A., F., & Udell, G. (1998). The economics of small business finance: The roles of private equity and debt markets in the financial growth cycle. *Journal of Banking & Finance*, *22*(6), 613–673. doi:10.1016/S0378-4266(98)00038-7

Bernick, M. (n.d.). *After Robots Take Over Our Jobs, Then What?* https://www.forbes.com/sites/michaelbernick/2017/04/11/after-robots-take-over-our-jobs-then-what

Bhatnagar, V., Sinha, S., Johri, P., & Bali, V. (2021). *Disruptive Technologies for Society 5.0: Exploration of New Ideas, Techniques, and Tools*. CRC Press.

Bhattacherjee, A. (2001). Understanding information systems continuance: An expectation-confirmation model. *Management Information Systems Quarterly*, *25*(3), 351–370. doi:10.2307/3250921

Blazhenkova, O., & Kozhevnikov, M. (2009). The New Object-Spatial-Verbal Cognitive Style Model: Theory and measurement. *Applied Cognitive Psychology*, *23*(5), 638–663. doi:10.1002/acp.1473

Block, D. (2004). Globalization, Transnational Communication, and the Internet. *International Journal on Multicultural Societies*, 6.

Blowers, M. (2015). *Evolution of cyber technologies and operations to 2035*. Springer. doi:10.1007/978-3-319-23585-1

Bolotova, Y. V. (2009). Cartel overcharges: An empirical analysis. *Journal of Economic Behavior & Organization*, *70*(1), 321–341. doi:10.1016/j.jebo.2009.02.002

Bonnet, F., & Venkatesh, S. (2016). *Poverty and Informal Economies*. Oxford University Press. https://halshs.archives-ouvertes.fr/halshs-01297260

Bradley, D. A., Burd, N., Dawson, D., & Loader, A. J. (2018). *Mechatronics: electronics in products and processes*. Routledge. doi:10.1201/9780203747735

Bradley, G. (2006). *Social and Community Informatics: Humans on the Net*. Routledge.

Brandon, E. (2012). *Global approaches to site contamination law*. Springer Science & Business Media.

Britton-Purdy, J., Grewal, D. S., Kapczynski, A., & Rahman, K. S. (2019). Building a Law-and-Political-Economy Framework: Beyond the Twentieth-Century Synthesis. *The Yale Law Journal*, *129*, 1784.

Brönneke, J. B., Müller, J., Mouratis, K., Hagen, J., & Stern, A. D. (2021). Regulatory, legal, and market aspects of intelligent wearables for cardiac monitoring. *Sensors (Basel)*, *21*(14), 4937. doi:10.339021144937 PMID:34300680

Brown, R. (2007). *Promoting entrepreneurship in arts education*. doi:10.4337/9781848440128.00017

Büchi, G., Cugno, M., & Castagnoli, R. (2018). Economies of Scale and Network Economies in Industry 4.0. Theoretical Analysis and Future Directions of Research. Symphonya. *Emerging Issues in Management*, *2*(2), 6. doi:10.4468/2018.02.01

Burgers, W. P., Hill, C. W. L., & Kim, W. C. (1993). A theory of global strategic alliances: The case of the global auto industry. *Strategic Management Journal*, *14*(6), 419–432. doi:10.1002mj.4250140603

Burns, D. J., & Neisner, L. (2006). Customer satisfaction in a retail setting: The contribution of emotion. *International Journal of Retail & Distribution Management*, *34*(1), 49–66. doi:10.1108/09590550610642819

Buss, D. D. (2002). Technology in the Internet age. *2002 IEEE International Solid-State Circuits Conference. Digest of Technical Papers (Cat. No.02CH37315),* (pp. 18–21). IEEE. 10.1109/ISSCC.2002.992920

Butler, D. L. (2002). Individualizing instruction in self-regulated learning. *Theory into Practice, 41*(2), 81–92. doi:10.120715430421tip4102_4

Cabinet Office. (2023). *Society 5.0.* http://www8.cao.go.jp/cstp/ english/society5_0/index.html

Callea, G., Federici, C., Freddi, R., & Tarricone, R. (2022). Recommendations for designing and implementing an Early Feasibility Studies program for medical devices in the European Union. *Expert Review of Medical Devices.*

Campos, N. F., Estrin, S., & Proto, E. (2010). *Corruption as a Barrier to Entry: Theory and Evidence (SSRN* Scholarly Paper ID 1711074). Social Science Research Network. https://papers.ssrn.com/abstract=1711074 doi:10.2139/ssrn.1693340

Cantor, R., & Hewlett, J. (1988). The economics of nuclear power: Further evidence on learning, economies of scale, and regulatory effects. *Resources and Energy, 10*(4), 315–335. doi:10.1016/0165-0572(88)90009-6

Carare, P. M. (2011). *Monopoly: Advantages and Disadvantages (SSRN* Scholarly Paper ID 1787089). Social Science Research Network. doi:10.2139/ssrn.1787089

Cep Ubad, A., Vina, A., & Ade Gafar, A. (2020). Borderless Education as a Challenge in the 5.0 Society. In *Proceedings of the 3rd International Conference on Educational Sciences (ICES 2019).* CRC Press.

Ch, M. A., Faheem, M. A., Dost, M. K. B., & Abdullah, I. (2011). Globalization and its *Impacts on the World Economic Development, 2*(23), 8.

Chandler, G. N., & Mcevoy, G. M. (2000). Human Resource Management, TQM, and Firm Performance in Small and Medium-Size Enterprises. *Entrepreneurship Theory and Practice, 25*(1), 43–58. doi:10.1177/104225870002500105

Changchun, S. W., Lee, C.-F., & Hsu, Y.-J. (2004). Online personalized sales promotion in electronic commerce. *Expert Systems with Applications, 27*(1), 35–52. doi:10.1016/j.eswa.2003.12.017

Charoenrat, T., Harvie, C., & Amornkitvikai, Y. (2013). Thai manufacturing small and medium-sized enterprise technical efficiency: Evidence from firm-level industrial census data. *Journal of Asian Economics, 27*, 42–56. doi:10.1016/j.asieco.2013.04.011

Chatfielda & Reddick. (2019). A framework for Internet of Things-enabled smart government: A case of IoTcybersecurity policies and use cases in U.S. federal government. *Government Information Quarterly, 36*(2), 346-357.

Chattopadhyay, S., Shankar, S., Gangadhar, R. B., & Kasinathan, K. (2018). Applications of artificial intelligence in assessment for learning in schools. In *Handbook of research on digital content, mobile learning, and technology integration models in teacher education* (pp. 185–206). IGI Global. doi:10.4018/978-1-5225-2953-8.ch010

Chen, D.-N., Hu, P. J.-H., Kuo, Y.-R., & Liang, T.-P. (2010). A Web-based personalized recommendation system for mobile phone selection: Design, implementation, and evaluation. *Expert Systems with Applications*, *37*(12), 8201–8210.

Chen, M., Saad, W., & Yin, C. (2019). Liquid state machine learning for resource and cache management in LTE-U unmanned aerial vehicle (UAV) networks. *IEEE Transactions on Wireless Communications*, *18*(3), 1504–1517. doi:10.1109/TWC.2019.2891629

Chen, T., Peng, L., Yin, X., Rong, J., Yang, J., & Cong, G. (2020, September). Analysis of user satisfaction with online education platforms in China during the COVID-19 pandemic. *Health Care*, *8*(3), 200. PMID:32645911

Chen, Y. J., Chiou, C. M., Huang, Y. W., Tu, P. W., Lee, Y. C., & Chien, C. H. (2018). A comparative study of medical device regulations: US, Europe, Canada, and Taiwan. *Therapeutic Innovation & Regulatory Science*, *52*(1), 62–69. doi:10.1177/2168479017716712 PMID:29714608

Chen, Z., Ling, K. C., Ying, G. X., & Meng, T. C. (2012). Antecedents of online customer satisfaction in China. *International Business Management*, *6*(2), 168–175. doi:10.3923/ibm.2012.168.175

Cheung, C. M. K., & Lee, M. K. O. (2012). What drives consumers to spread electronic word of mouth in online consumer-opinion platforms. *Decision Support Systems*, *53*(1), 21825. doi:10.1016/j.dss.2012.01.015

Chew, S. C., Robertson, R., & Garrett, W. R. (1993). Review of Globalization: Social Theory and Global Culture.; Religion and Global Order., Roland Robertson [Review of Review of Globalization: Social Theory and Global Culture.; Religion and Global Order., Roland Robertson, by R. Robertson & W. R. Garrett]. *Contemporary Sociology*, *22*(6), 828–830. doi:10.2307/2075975

Cho, Y. H., Kim, J. K., & Kim, S. H. (2002). A personalized recommender system based on web usage mining and decision tree induction. *Expert Systems with Applications*, *23*(3), 329–342. doi:10.1016/S0957-4174(02)00052-0

Chung, K., & Shin, J. (2010). The antecedents and consequences of relationship quality in internet shopping. *Asia Pacific Journal of Marketing and Logistics*, *22*(4), 473–491. doi:10.1108/13555851011090510

Coe, D. T. (2007). *Globalization and Labour Markets: Policy Issues Arising from the Emergence of China and India*. OECD. doi:10.1787/1815199X

Compton, R. A., Giedeman, D. C., & Hoover, G. A. (2011). Panel evidence on economic freedom and growth in the United States. *European Journal of Political Economy*, *27*(3), 423–435. doi:10.1016/j.ejpoleco.2011.01.001

Conderman, G., Ikan, P. A., & Hatcher, R. E. (2000). Student-led conferences in inclusive settings. *Intervention in School and Clinic*, *36*(1), 22–26. doi:10.1177/105345120003600103

Coronado, E., Kiyokawa, T., Ricardez, G. A. G., Ramirez-Alpizar, I. G., Venture, G., & Yamanobe, N. (2022). Evaluating quality in human-robot interaction: A systematic search and classification of performance and human-centered factors, measures and metrics towards an industry 5.0. *Journal of Manufacturing Systems*, *63*, 392–410. doi:10.1016/j.jmsy.2022.04.007

Cortright, J. (2002). The Economic Importance of Being Different: Regional Variations in Tastes, Increasing Returns, and the Dynamics of Development. *Economic Development Quarterly*, *16*(1), 3–16. doi:10.1177/0891242402016001001

Council, B. (2018). *Future skills supporting the UAE's future workforce*. British Council. www. Britishcouncil. Ae/Sites/Default/Files/Bc_futureskills_english_1mar18_3. pdf

Covington, P., Adams, J., & Sargin, E. (2016). Deep neural networks for youtube recommendations. *Proceedings of the 10th ACM conference on recommender systems*, 191–198. 10.1145/2959100.2959190

Crafts, N. (2004). Globalization and Economic Growth: A Historical Perspective. *World Economy*, *27*(1), 45–58. doi:10.1111/j.1467-9701.2004.00587.x

Cui, Z., Xu, X., Fei, X. U. E., Cai, X., Cao, Y., Zhang, W., & Chen, J. (2020). Personalized recommendation system based on collaborative filtering for IoT scenarios. *IEEE Transactions on Services Computing*, *13*(4), 685–695. doi:10.1109/TSC.2020.2964552

Cunningham, P., Bergmann, R., Schmitt, S., Traphoner, R., Breen, S., & Smyth, B. (2001). WebSell: Intelligent sales assistants for the World Wide Web. *Kunstliche Intelligenz*, *15*(1), 28–32.

Dash, G., Kiefer, K., & Paul, J. (2021). Marketing-to-Millennials: Marketing 4.0, customer satisfaction and purchase intention. *Journal of Business Research*, *122*, 608–620. doi:10.1016/j.jbusres.2020.10.016

Davenport, T., Guha, A., Grewal, D., & Bressgott, T. (2020). How artificial intelligence will change the future of marketing. *Journal of the Academy of Marketing Science*, *48*(1), 24–42. doi:10.100711747-019-00696-0

David, A., & Barwinska-Malajowicz, A. (2018). Where do We Go from Here? The EU Migration Flows after the Brexit Referendum. Possible Future Scenarios in the Polish Example. *Journal of Globalization Studies*, *9*(2), 3–17. doi:10.30884/jogs/2018.02.01

Dawoud, D., Naci, H., Ciani, O., & Bujkiewicz, S. (2021). Raising the bar for using surrogate endpoints in drug regulation and health technology assessment. *BMJ (Clinical Research Ed.)*, 374. doi:10.1136/bmj.n2191 PMID:34526320

Dawson, P., Henderson, M., Ryan, T., Mahoney, P., Boud, D., Phillips, M., & Molloy, E. (2018). Technology and feedback design. *Learning, design, and technology*.

de Morais Watanabe, E. A., Torres, C. V., & Alfinito, S. (2019). *The impact of culture, evaluation of store image and satisfaction on purchase intention at supermarkets*. Revista de Gestão.

Dhingra, S., Gupta, S., & Bhatt, R. (2020). A study of relationship among service quality of E-commerce websites, customer satisfaction, and purchase intention. *International Journal of E-Business Research*, *16*(3), 42–59. doi:10.4018/IJEBR.2020070103

Di Nardo, M., & Yu, H. (2021). Special Issue "Industry 5.0: The Prelude to the Sixth Industrial Revolution". *Applied System Innovation*, *4*(3), 45. doi:10.3390/asi4030045

Diaconu, K., Chen, Y. F., Cummins, C., Jimenez Moyao, G., Manaseki-Holland, S., & Lilford, R. (2017). Methods for medical device and equipment procurement and prioritization within low-and middle-income countries: A systematic literature review findings. *Globalization and Health*, *13*(1), 1–16. doi:10.118612992-017-0280-2 PMID:28821280

Dianat, I., Adeli, P., Jafarabadi, M. A., & Karimi, M. A. (2019). User-centred web design, usability and user satisfaction: The case of online banking websites in Iran. *Applied Ergonomics*, *81*, 102892. doi:10.1016/j.apergo.2019.102892 PMID:31422242

Djolov, G. G. (2014). The Economics of Competition. *The Race to Monopoly*. Taylor and Francis. https://www.taylorfrancis.com/books/mono/10.4324/9781315785561/economics-competition-george-djolov

Domonoske, C. (2019, Apr. 30). *Even in the Robot Age, Manufacturers Need the Human Touch*. NPR.

Dostaler, I., & Tomberlin, J. (2013). The great divide between business schools research and business practice. *Canadian Journal of Higher Education*, *43*(1), 115–128. doi:10.47678/cjhe.v43i1.1895

Doyle-Kent, M., & Kopacek, P. (2019). Industry 5.0: Is the Manufacturing Industry on the Cusp of a New Revolution? In *Proceedings of the International Symposium for Production Research* (pp. 432-441). Springer.

Doyle-Kent, M., & Shanahan, B. W. (2022). The development of a novel educational model to successfully upskill technical workers for Industry 5.0: Ireland a case study. *IFAC-PapersOnLine*, *55*(39), 425–430. doi:10.1016/j.ifacol.2022.12.072

Dreher, A., Gaston, N., & Martens, P. (2008). *Measuring Globalization*. Springer New York., doi:10.1007/978-0-387-74069-0

Dutta, S., Pramanik, S., & Bandyopadhyay, S. K. (2021). Prediction of Weight Gain during COVID-19 for Avoiding Complication in Health. *International Journal of Medical Science and Current Research*, *4*(3), 1042–1052.

Ehlers, U.-D., & Kellermann, S. A. (2019). Future skills: The future of learning and higher education. Karlsruhe.

Eisenman, A., Naumov, M., Gardner, D., Smelyanskiy, M., Pupyrev, S., Hazelwood, K., Cidon, A., & Katti, S. (2018). *Bandana: Using nonvolatile memory for storing deep learning models.* arXiv preprint arXiv:1811.05922.

Eklof, J., Podkorytova, O., & Malova, A. (2020). Linking customer satisfaction with financial performance: An empirical study of Scandinavian banks. *Total Quality Management & Business Excellence*, *31*(15-16), 1684–1702. doi:10.1080/14783363.2018.1504621

El-Adly, M. I. (2019). Modelling the relationship between hotel perceived value, customer satisfaction, and customer loyalty. *Journal of Retailing and Consumer Services*, *50*, 322–332. doi:10.1016/j.jretconser.2018.07.007

Elangovan, U. (2021). *Industry 5.0: The Future of the Industrial Economy.* CRC. doi:10.1201/9781003190677

Elavarasan, R. M., Leoponraj, S., Vishnupriyan, J., Dheeraj, A., & Sundar, G. G. (2021). Multi-Criteria Decision Analysis for user satisfaction-induced demand-side load management for an institutional building. *Renewable Energy*, *170*, 1396–1426. doi:10.1016/j.renene.2021.01.134

EmiratesSkills. (2023). https://www.emiratesskills.ae

Ershova, N. (2011). *Medical device regulations as a source of industrial leadership: A comparative study of American and European regulatory approaches.* Academic Press.

EzumaM.ErdenF.AnjinappaC. K.OzdemirO.GuvencI. (2019). Micro-UAV Detection and Classification from RF Fingerprints Using Machine Learning Techniques. doi:10.1109/AERO.2019.8741970

Farshadkhah, S., Van Slyke, C., & Fuller, B. (2021). Onlooker effect and affective responses in information security violation mitigation. *Computers & Security*, *100*, 102082. doi:10.1016/j.cose.2020.102082

Farzanegan, M. R., Feizi, M., & Gholipour, H. F. (2021). Globalization and the Outbreak of COVID-19: An Empirical Analysis. *Journal of Risk and Financial Management*, *14*(3), 105. doi:10.3390/jrfm14030105

Feldman, J. M., & Lynch, J. G. (1988). Self-generated validity and other effects of measurement on belief, attitude, intention and behavior. *The Journal of Applied Psychology*, *73*(3), 42135. doi:10.1037/0021-9010.73.3.421

Fink, L. (2019). *Purpose & profit.* Harvard University. https://corpgov.law.harvard.edu/2019/01/23/purpose-profit/

Foray, D., Mowery, D. C., & Nelson, R. R. (2012). Public R&D; and social challenges: What lessons from mission R&D; programs? *Research Policy*, *41*, 1697-1702.

Fornell, C. (1992). A national customer satisfaction barometer: The Swedish experience. *Journal of Marketing*, *56*(1), 6–21. doi:10.1177/002224299205600103

Future Skills. (2022). *Future Skills -Kingston University-UK.* https://www.kingston.ac.uk/aboutkingstonuniversity/future-skills/

Fu, Y., Wan, J., Zhao, H., Jiang, W., & Pu, S. (2020). Preference-aware heterogeneous graph neural networks for recommendation. *2020 IEEE 32nd international conference on tools with artificial intelligence (ICTAI)*, 41–46. 10.1109/ICTAI50040.2020.00017

García-Sánchez, I. M., & García-Sánchez, A. (2020). Corporate social responsibility during the COVID-19 pandemic. *Journal of Open Innovation*, 6(4), 126. doi:10.3390/joitmc6040126

Garmann-Johnsen, N. F., & Eikebrokk, T. R. (2017). Dynamic capabilities in e-health innovation: Implications for policies. *Health Policy and Technology*, 6(3), 292–301. doi:10.1016/j.hlpt.2017.02.003

GBD 2015 SDG Collaborators. (2016). Measuring the health-related Sustainable Development Goals in 188 countries: A baseline analysis from the Global Burden of Disease Study 2015. *Lancet, 388*(10053), 1813–1850. doi:10.1016/S0140-6736(16)31467-2

Ge, X., Wang, J., Ding, J., Cao, X., Zhang, Z., Liu, J., & Li, X. (2019). Combining UAV-based hyperspectral imagery and machine learning algorithms for soil moisture content monitoring. *PeerJ, 7*, e6926. doi:10.7717/peerj.6926 PMID:31110930

Gezim, J., & Bashkim, B. (2019). Trade Barriers and Exports between Western Balkan Countries. *Naše Gospodarstvo/Our Economy, 65*(4), 72–80.

Ghasemaghaei, M., Hassanein, K., & Benbasat, I. (2019). Assessing the design choices for online recommendation agents for older adults: Older does not always mean simpler information technology. *Management Information Systems Quarterly*, 43(1), 329–346. doi:10.25300/MISQ/2019/13947

Glaeser, E. L., Laibson, D. I., Scheinkman, J. A., & Soutter, C. L. (2000). Measuring Trust. *The Quarterly Journal of Economics*, 115(3), 811–846. doi:10.1162/003355300554926

Goldstein, N. (2010). *Globalization and Free Trade.* Infobase Publishing.

Gomez-Uribe, C. A., & Hunt, N. (2015). The Netflix recommender system: Algorithms, business value, and innovation. *ACM Transactions on Management Information Systems*, 6(4), 1–19. doi:10.1145/2843948

Gonzo, R., & Pokraka, A. (2021). Light-ray operators, detectors and gravitational event shapes. *J. High Energy. Phys., 15*. doi:10.1007/JHEP05(2021)015

Goodreads. (2022). *Quotable Quote.* Goodreads. https://www.goodreads.com/quotes/556030-imagination-is-more-important-than-knowledge-for-knowledge-is-limited

Greenwell, T. C., Fink, J. S., & Pastore, D. L. (2002). Assessing the influence of the physical sports facility on customer satisfaction within the context of the service experience. *Sport Management Review*, 5(2), 129–148. doi:10.1016/S1441-3523(02)70064-8

Greeven, M. J., Yip, G. S., & Wei, W. (2019). *Pioneers, Hidden Champions, Changemakers, and Underdogs: Lessons from China's Innovators.* MIT Press. doi:10.7551/mitpress/12007.001.0001

Griffiths, J. R., Johnson, F., & Hartley, R. J. (2007). User satisfaction as a measure of system performance. *Journal of Librarianship and Information Science, 39*(3), 142–152. doi:10.1177/0961000607080417

Grönroos, C. (1994). From scientific management to service management: A management perspective for the age of service competition. *International Journal of Service Industry Management, 5*(1), 5–20. doi:10.1108/09564239410051885

Güğerçin, S. (2021). How Employees Survive In The Industry 5.0 Era: In-Demand Skills Of The Near Future. Int. J. Discip. Econ. Adm. Sci. Stud. *IDEAstudies, 7*(31), 524–533. doi:10.26728/ideas.452

Guo, X., Wang, J., Zhao, W., Zhang, K., & Wang, C. (2016). Study medical device innovation design strategy based on demand analysis and process case base. *Multimedia Tools and Applications, 75*(22), 14351–14365. doi:10.100711042-015-3176-2

Gupta, U., Wu, C. J., Wang, X., Naumov, M., Reagen, B., Brooks, D., ... Lee, H. H. S. (2020, February). The architectural implications of facebook's DNN-based personalized recommendation. In *2020 IEEE International Symposium on High Performance Computer Architecture (HPCA)* (pp. 488-501). IEEE. 10.1109/HPCA47549.2020.00047

Gygli, S., Haelg, F., Potrafke, N., & Sturm, J.-E. (2019). The KOF Globalisation Index – revisited. *The Review of International Organizations, 14*(3), 543–574. doi:10.100711558-019-09344-2

Hadidi, H. E., & Kirby, D. A. (2015). Universities and Innovation in a Factor-Driven Economy: The Egyptian Case. *Industry and Higher Education, 29*(2), 151–160. doi:10.5367/ihe.2015.0248

Hahn, G. J. (2020). Industry 4.0: A supply chain innovation perspective. *International Journal of Production Research, 58*(5), 1425–1441. doi:10.1080/00207543.2019.1641642

Hallion, A. M. (1994). *Strategies for Developing Multi-Age Classrooms.* Academic Press.

Harel, Y., Gal, I. B., & Elovici, Y. (2017). *Cyber security and the role of intelligent systems in addressing its challenges* (Vol. 8). ACM New York.

Harkins, M. W. (2016). *Managing risk and information security: protect to enable.* Springer Nature. doi:10.1007/978-1-4842-1455-8

Hax, A. C. (1989). *Building The Firm Of The Future.* ProQuest. https://www.proquest.com/openview/63c2ebf36830d8c46ee32fc7482c4ecc/1?pq-origsite=gscholar&cbl=26142

Heimberger, P. (2020). Does economic globalization affect income inequality? A meta-analysis. *World Economy, 43*(11), 2960–2982. doi:10.1111/twec.13007

Henderson, R. (2021). *Reimagining Capitalism in a World on Fire: Shortlisted for the FT & McKinsey Business Book of the Year Award 2020.* Penguin UK.

Hennipman, P. (1954). Monopoly: Impediment or Stimulus to Economic Progress? In E. H. Chamberlin (Ed.), *Monopoly and Competition and their Regulation: Papers and Proceedings of a Conference held by the International Economic Association* (pp. 421–456). Palgrave Macmillan UK. 10.1007/978-1-349-08434-0_22

Herlocker, J. L., Konstan, J. A., Terveen, L. G., & Riedl, J. T. (2004). Evaluating Collaborative Filtering Recommender Systems. *ACM Transactions on Information Systems*, *22*(1), 5–53. doi:10.1145/963770.963772

Hewitt, T., & Wield, D. (1992). In T. Hewitt, D. Wield, & H. Johnson (Eds.), *Technology and Industrialization* (pp. 201–221). Oxford University Press. https://www.amazon.co.uk/Industrialization-Development-Tom-Hewitt/dp/0198773323

Hoang, H., & Antoncic, B. (2003). Network-based research in entrepreneurship: A critical review. *Journal of Business Venturing*, *18*(2), 165–187. doi:10.1016/S0883-9026(02)00081-2

Holbrook, T. (1996). *Do campaigns matter?* (Vol. 1). Sage Publications. doi:10.4135/9781452243825

Hossain, M. S., Zhou, X., & Rahman, M. F. (2018). Examining the impact of QR codes on purchase intention and customer satisfaction on the basis of perceived flow. *International Journal of Engineering Business Management*, *10*. doi:10.1177/1847979018812323

Hsu, C. L., & Lin, J. C. C. (2015). What drives purchase intention for paid mobile apps?–An expectation confirmation model with perceived value. *Electronic Commerce Research and Applications*, *14*(1), 46–57. doi:10.1016/j.elerap.2014.11.003

Huang, Y., Wang, N., Zhang, H., & Wang, J. (2019). A novel product recommendation model consolidating price, trust and online reviews. *Kybernetes*, *48*(6), 1355–1372. doi:10.1108/K-03-2018-0143

Hui, B., Zhang, L., Zhou, X., Wen, X., & Nian, Y. (2022). Personalized recommendation system based on knowledge embedding and historical behavior. *Applied Intelligence*, *52*(1), 954–966. doi:10.100710489-021-02363-w

Hunt, R. s. (2018). Learning Data Analytics Carpenteria, CA, linkedin.com.

Huynh, T. D., Ebden, M., Fischer, J., Roberts, S., & Moreau, L. (2018). Provenance Network Analytics: An approach to data analytics using data provenance. *Data Mining and Knowledge Discovery*, *32*(3), 708–735. doi:10.100710618-017-0549-3

IBE-UNESCO. (2017). *Training Tools for Curriculum Development: Personalized Learning.* https://unesdoc.unesco.org/ark:/48223/pf0000250057/PDF/250057eng.pdf.multi

IGI Global Dictionary. (2023). *What is Skill.* https://www.igi-global.com/dictionary/combining-local-global-expertise-services/27088

International Conference on Educational. (2020). Borderless education as a challenge in the 5.0 society. In *proceedings of the 3rd International Conference on Educational Sciences (ICES 2019).* Routledge.

Ismail, M., Shahin, M., Shaaban, M. F., Serpedin, E., & Qaraqe, K. (2018). Efficient detection of electricity theft cyber attacks in AMI networks. Paper presented at the *2018 IEEE Wireless Communications and Networking Conference (WCNC)*. IEEE. 10.1109/WCNC.2018.8377010

Iyengar, K. P., Pe, E. Z., Jalli, J., Shashidhara, M. K., Jain, V. K., Vaish, A., & Vaishya, R. (2022). Industry 5.0 technology capabilities in Trauma and Orthopaedics. *Journal of Orthopaedics*, *32*, 125–132. doi:10.1016/j.jor.2022.06.001 PMID:35707297

Jacks, D. S., & Novy, D. (2020). Trade Blocs and Trade Wars during the Interwar Period. *Asian Economic Policy Review*, *15*(1), 119–136. doi:10.1111/aepr.12276

James, H. (2009). The End of Globalization: Lessons from the Great Depression. Harvard University Press.

Jamil, M. I. M., & Almunawar, M. N. (2021). Importance of Digital Literacy and Hindrance Brought About by Digital Divide [Chapter]. Encyclopedia of Information Science and Technology, Fifth Edition. IGI Global. doi:10.4018/978-1-7998-3479-3.ch116

Javaid, M., & Haleem, A. (2020). Critical components of Industry 5.0 towards a successful adoption in the field of manufacturing. *Journal of Industrial Integration and Management*, *5*(3), 327–348. doi:10.1142/S2424862220500141

Jiang, Z., & Benbasat, I. (2004). Virtual product experience: Effects of visual and functional control of products on perceived diagnosticity in electronic shopping. *Journal of Management Information Systems*, *21*(3), 111–147. doi:10.1080/07421222.2004.11045817

Jimeno Muñoz, J. (2019). Cyber Risks: Liability and Insurance. The Extraordinary Risks in a Hyperconnectivity World. *InDret, 2*.

Jnr, B. A. (2020). Examining the role of green IT/IS innovation in collaborative enterprise-implications in an emerging economy. *Technology in Society*, *62*, 101301.

Jónsson, S. (2019). *RGB and Multispectral UAV image classification of agricultural fields using a machine learning algorithm*. Student thesis series INES.

Joseph, S. (1999). Taming the Leviathans: Multinational Enterprises and Human Rights. *Netherlands International Law Review*, *46*(2), 171–203. doi:10.1017/S0165070X00002394

Karmarkar, U. R., Shiv, B., & Knutson, B. (2015). Cost conscious? The neural and behavioral impact of price primacy on decision-making. *JMR, Journal of Marketing Research*, *52*(4), 467–481. doi:10.1509/jmr.13.0488

Kaushik, D., Garg, M., Annu, Gupta, A., & Pramanik, S. (2021). Application of Machine Learning and Deep Learning in Cyber security: An Innovative Approach. InGhonge, M., Pramanik, S., Mangrulkar, R., & Le, D. N. (Eds.), *Cybersecurity and Digital Forensics: Challenges and Future Trends*. Wiley.

Khairunniza-Bejo, S., Mustaffha, S., & Wan, I. W. I. (2014). Application of artificial neural network in predicting crop yield: A review. *Journal of Food Science and Engineering*, *4*(1), 1.

Khanna, R., Guler, I., & Nerkar, A. (2016). Fail often, fail big, and fail fast? Learning from small failures and R&D performance in the pharmaceutical industry. *Academy of Management Journal*, *59*(2), 436–459. doi:10.5465/amj.2013.1109

Khosravi, V., Ardejani, F. D., Yousefi, S., & Aryafar, A. (2018). Monitoring soil lead and zinc contents via combination of spectroscopy with extreme learning machine and other data mining methods. *Geoderma*, *318*, 29–41. doi:10.1016/j.geoderma.2017.12.025

Kim, H.-W., Xu, Y., & Gupta, S. (2012). Which is more important in internet shopping, perceived price or trust. *Electronic Commerce Research and Applications*, *11*(3), 241–252. doi:10.1016/j.elerap.2011.06.003

Kim, J.-K., Choi, I.-Y., & Li, Q. (2021). Customer Satisfaction of Recommender System: Examining Accuracy and Diversity in Several Type Recommendation Approach. *Sustainability (Basel)*, *13*(11), 6165. doi:10.3390u13116165

Krasniqi, B. (2007). Barriers to entrepreneurship and SME growth in transition: The case of Kosova. [JDE]. *Journal of Developmental Entrepreneurship*, *12*(1), 71–94. doi:10.1142/S1084946707000563

Kudina, A., Yip, G., & Barkema, H. (2008). Born Global. *Business Strategy Review*, *19*(4), 38–44. doi:10.1111/j.1467-8616.2008.00562.x

Kumar, S., & Mallipeddi, R. R. (2022). Impact of cybersecurity on operations and supply chain management: Emerging trends and future research directions. *Production and Operations Management*, *31*(12), 4488–4500. doi:10.1111/poms.13859

Kurt, M. N., Ogundijo, O., Li, C., & Wang, X. (2018). Online cyber-attack detection in smart grid: A reinforcement learning approach. *IEEE Transactions on Smart Grid*, *10*(5), 5174–5185. doi:10.1109/TSG.2018.2878570

Kushi, L. H., Doyle, C., McCullough, M., Rock, C. L., Demark-Wahnefried, W., Bandera, E. V., Gapstur, S., Patel, A. V., Andrews, K., & Gansler, T. (2012). American Cancer Society Guidelines on nutrition and physical activity for cancer prevention: Reducing cancer risk with healthy food choices and physical activity. *CA: a Cancer Journal for Clinicians*, *62*(1), 30–67. doi:10.3322/caac.20140 PMID:22237782

Kwak, G.-H., & Park, N.-W. (2019). Impact of Texture Information on Crop Classification with Machine Learning and UAV Images. *Applied Sciences (Basel, Switzerland)*, *9*(4), 643. doi:10.3390/app9040643

Kwan, A., Morris, J., & Barbic, S. (2021). 811Protocol: A mixed methods evaluation of an IPS program to increase employment and well-being for people with long-term experience of complex barriers in Vancouver's downtown and DTES. *PLoS One*, *16*(12), e0261415. PMID:34914771

Kwon, S. (2009). Thirty years of national health insurance in South Korea: Lessons for achieving universal health care coverage. *Health Policy and Planning*, *24*(1), 63–71. doi:10.1093/heapol/czn037 PMID:19004861

Kwon, Y., Park, J., & Son, J.-Y. (2021). Accurately or accidentally? Recommendation agent and search experience in over-the-top (OTT) services. *Internet Research*, *31*(2), 562–586. doi:10.1108/INTR-03-2020-0127

Kwon, Y., Son, S., & Jang, K. (2020). User satisfaction with battery electric vehicles in South Korea. *Transportation Research Part D, Transport and Environment*, *82*, 102306. doi:10.1016/j.trd.2020.102306

Lantada, A. D. (2022). Engineering education 5.0: Strategies for a successful transformative project-based learning. *Insights Into Global Engineering Education After the Birth of Industry 5.0*, 19.

Laszlo, C., Cooperrider, D., & Fry, R. (2020). Global challenges as opportunity to transform business for good. *Sustainability (Basel)*, *12*(19), 8053. doi:10.3390u12198053

Lavorgna, L., Iaffaldano, P., Abbadessa, G., Lanzillo, R., Esposito, S., Ippolito, D., Sparaco, M., Cepparulo, S., Lus, G., Viterbo, R., Clerico, M., Trojsi, F., Ragonese, P., Borriello, G., Signoriello, E., Palladino, R., Moccia, M., Brigo, F., Troiano, M., ... Bonavita, S. (2022). Disability assessment using Google Maps. *Neurological Sciences*, *43*(2), 1007–1014. doi:10.100710072-021-05389-7 PMID:34142263

Learning Compass. (2018). *The OECD Learning Compass 2030*. OECD. https://www.oecd.org/education/2030-project/teaching-and-learning/learning/

Lee, E., & Vivarelli, M. (2006). The social impact of globalization in the developing countries. *International Labour Review*, *145*(3), 167–184. doi:10.1111/j.1564-913X.2006.tb00016.x

Lee, J., & Whaley, J. E. (2019). Determinants of dining satisfaction. *Journal of Hospitality Marketing & Management*, *28*(3), 351–378. doi:10.1080/19368623.2019.1523031

Lee, M.-C. (2010). Explaining and predicting users' continuance intention toward e-learning: An extension of the expectation–confirmation model. *Computers & Education*, *54*(2), 506–516. doi:10.1016/j.compedu.2009.09.002

Lee, M., Yun, J. J., Pyka, A., Won, D., Kodama, F., Schiuma, G., Park, H. S., Jeon, J., Park, K. B., Jung, K. H., Yan, M.-R., Lee, S. Y., & Zhao, X. (2018). How to respond to the fourth industrial revolution or the second information technology revolution? Dynamic new combinations between technology, market, and Society through open innovation. *Journal of Open Innovation*, *4*(3), 21. doi:10.3390/joitmc4030021

Lee, S. J., & Lee, H. C. (2007). A Study on Prediction Performance of Correspondence Average Algorithm in Cooperative Filtering Recommendation. *InformationSystem Review*, *9*(1), 85–103.

Lee, Y., Lin, B.-W., Wong, Y.-Y., & Calantone, R. J. (2011). Understanding and Managing International Product Launch: A Comparison between Developed and Emerging Markets: Understanding And Managing International Product Launch. *Journal of Product Innovation Management*, *28*(s1), 104–120. doi:10.1111/j.1540-5885.2011.00864.x

Lettl, C., Herstatt, C., & Gemuenden, H. G. (2006). Users' contributions to radical innovation: Evidence from four cases in the field of medical equipment technology. *Research Management*, *36*(3), 251–272.

Lewis, R. C., & Booms, B. H. (1983). The Marketing Aspects of Service Quality. In L. Berry, L. Shostack, & G. Upah (Eds.), *Emerging Perspectives on Services Marketing* (pp. 99–107). American Marketing Association.

Lezzi, M., Lazoi, M., & Corallo, A. (2018). Cybersecurity for Industry 4.0 in the current literature: A reference framework. *Computers in Industry*, *103*, 97–110. doi:10.1016/j.compind.2018.09.004

Liang, T., Lai, H., & Ku, Y. (2007). Personalized Content Recommendation and User Satisfaction: Theoretical Synthesis and Empirical Findings. *Journal of Management Information Systems*, *23*(3), 45–70. doi:10.2753/MIS0742-1222230303

Lin, H., Zhang, M., & Gursoy, D. (2020). Impact of nonverbal customer-to-customer interactions on customer satisfaction and loyalty intentions. *International Journal of Contemporary Hospitality Management*, *32*(5), 1967–1985. doi:10.1108/IJCHM-08-2019-0694

Liu, X., Liu, Y., Chen, Y., & Hanzo, L. (2018). *Trajectory design and power control for multi-UAV assisted wireless networks: a machine learning approach.* arXiv preprint arXiv:1812.07665.

Liu, C., Li, H., Tang, Y., Lin, D., & Liu, J. (2019). Next generation integrated smart manufacturing based on big data analytics, reinforced learning, and optimal routes planning methods. *International Journal of Computer Integrated Manufacturing*, *32*(9), 820–831. doi:10.1080/09 51192X.2019.1636412

Liu, L., Cheung, C. M. K., & Lee, M. K. O. (2016). An empirical investigation of information sharing behavior on social commerce sites. *International Journal of Information Management*, *36*(5), 686–699. doi:10.1016/j.ijinfomgt.2016.03.013

Liu, S., Liu, X., Wang, S., & Muhammad, K. (2021). Fuzzy-aided solution for an out-of-view challenge in visual tracking under IoT-assisted complex environment. *Neural Computing & Applications*, *33*(4), 1055–1065. doi:10.100700521-020-05021-3

Liu, Y., & Jang, S. S. (2009). Perceptions of Chinese restaurants in the US: What affects customer satisfaction and behavioral intentions? *International Journal of Hospitality Management*, *28*(3), 338–348. doi:10.1016/j.ijhm.2008.10.008

Loretta, L., & Bang, S. (2015). Hyun, & Morales [Global Gentrifications: Uneven Development and Displacement. Policy Press.]. *E (Norwalk, Conn.)*, L.

Lumpkin, G. T., & Dess, G. G. (1996). Clarifying the Entrepreneurial Orientation Construct and Linking It To Performance. *Academy of Management Review*, *21*(1), 135–172. doi:10.2307/258632

Lu, Y., Zheng, H., Chand, S., Xia, W., Liu, Z., Xu, X., Wang, L., Qin, Z., & Bao, J. (2022). Outlook on human-centric manufacturing towards Industry 5.0. *Journal of Manufacturing Systems*, *62*, 612–627. doi:10.1016/j.jmsy.2022.02.001

Lynch, M. (2018, January 2). *How to Humanize the Education Machine*. https://www.edweek. org/education/opinion-how-to-humanize-the-education-machine/2018/01

Maddikunta, P. K. R., Pham, Q. V., Prabadevi, B., Deepa, N., Dev, K., Gadekallu, T. R., ... Liyanage, M. (2022). Industry 5.0: A survey on enabling technologies and potential applications. *Journal of Industrial Information Integration, 26*, 100257. doi:10.1016/j.jii.2021.100257

Máliková, I. (2020). Impact of globalization on circular economy and sustainable development. *SHS Web of Conferences, 74*, 06018. 10.1051hsconf/20207406018

Ma, N., Zeng, Z., Wang, Y., & Xu, J. (2021). Balanced strategy based on environment and user benefit-oriented carpooling service mode for commuting trips. *Transportation, 48*(3), 1241–1266. doi:10.100711116-020-10093-0

Mandloi & Inada. (2019). Machine Learning Approach for Drone Perception and Control. In *International Conference on Engineering Applications of Neural Networks*. Springer.

Marr, B. (2016). *Big Data in Practice: How 45 Successful Companies Used Big Data Analytics to Deliver Extraordinary Results*. John Wiley & Sons, Ltd. doi:10.1002/9781119278825

Martynov, V. V., Shavaleeva, D. N., & Zaytseva, A. A. (2019). Information Technology as the Basis for Transformation into a Digital Society and Industry 5.0. In *2019 International Conference Quality Management, Transport and Information Security, Information Technologies (IT&QM&IS)* (pp. 539-543). IEEE.

Maslowska, E., Smit, E. G., & van den Putte, B. (2016). It is all in the name: A study of consumers' responses to personalized communication. *Journal of Interactive Advertising, 16*(1), 74–85. do i:10.1080/15252019.2016.1161568

Mason, J., & Lefrere, P. (2003). Trust, Collaboration, e-Learning and Organisational Transformation. *International Journal of Training and Development, 7*(4), 259–270.

Mayer, J. (2002). Globalization, technology transfer, and skill accumulation in low-income countries. In *Globalization, Marginalization and Development*. Routledge. doi:10.4324/9780203427637.ch5

Mbugua, J. K., Mbugua, S. N., Wangoi, M., & Kariuki, J. O. O. & J. N. (2013). *Factors Affecting the Growth of Micro and Small Enterprises: A Case of Tailoring and Dressmaking Enterprises in Eldoret*. http://localhost:8080/xmlui/handle/123456789/9928

McKay, E. (2008). The Human-Dimensions of Human-Computer Interaction: Balancing the HCI Equation (1 ed., Vol. 3). IOS Press.

McKay, E., & Izard, J. (2013). Seamless Web-Mediated Training Courseware Design Model: Innovating Adaptive Educational-Learning Systems. In A. P. Ayala (Ed.), *Intelligent and Adaptive Educational-Learning Systems: Achievements and Trends, ISSN:2190-3018 (Vol. 17*, pp. 417-442). Springer Berlin Heidelberg.

McKay, E. (2008). *The Human-Dimensions of Human-Computer Interaction: Balancing the HCI Equation*. IOS Press.

McKay, E. (2016). Gearing up the knowledge engineers: Experience design through effective human-computer interaction (HCI). In A. Lugmayr & C. Dal-Zotto (Eds.), *Media Convergence Handbook* (Vol. 2, pp. 283–307). Springer-Verlag. doi:10.1007/978-3-642-54487-3_15

McKay, E. (2018). Prescriptive Training Courseware: IS-Design Methodology. *AJIS. Australasian Journal of Information Systems*, *22*. https://doi.org/http://dx.doi.org/10.3127/ajis.v22i0.1675

McKay, E., Asquith, K., Smyrnova-Trybulska, E., Porczyńska-Ciszewska, A., & Kopczyński, T. (2022). Data Evaluation of Happiness Scale Online Study: A Rasch Measurement Analysis. In E. McKay (Ed.), *Manage Your Own Learning Analytics: Implement a Rasch Modelling Approach* (Vol. 261, pp. 73–112). Springer. doi.org/10.1007/978-3-030-86316-6

McKay, E., & Mohamad, M. (2018). Big Data Management Skills: Accurate Measurement. *Research and Practice in Technology Enhanced Learning*, *13*(5). doi:10.118641039-018-0071-2 PMID:30595736

Medina, J. F., & Duffy, M. F. (1998). Standardization vs globalization: A new perspective of brand strategies. *Journal of Product and Brand Management*, *7*(3), 223–243. doi:10.1108/10610429810222859

Melvin, T., & Torre, M. (2019). New medical device regulations: The regulator's view. *Effort Open Reviews*, *4*(6), 351–356. doi:10.1302/2058-5241.4.180061 PMID:31312522

Merrill, M. D., Tennyson, R. D., & Posey, L. O. (1992). *Teaching Concepts: An instructional design guide* (2nd ed.). Educational Technology Publications.

Meuer, J., Rupietta, C., & Backes-Gellner, U. (2015). Layers of coexisting innovation systems. *Research Policy*, *44*(4), 888–910. doi:10.1016/j.respol.2015.01.013

Miller, C. C. (n.d.). *Evidence That Robots Are Winning the Race for American Jobs*. https://www.nytimes.com/2017/03/28/upshot/evidence-that-robots-are-winning-the-race-for-american-jobs.html

Miller, R. (2019). *Superb AI generates customized training data for machine learning projects*. TechCrunch.

Mittal, B., & Lassar, W. M. (1996). The role of personalization in service encounters. *Journal of Retailing*, *72*(1), 95–109. doi:10.1016/S0022-4359(96)90007-X

Mokhova, N., & Zinecker, M. (2014). Macroeconomic Factors and Corporate Capital Structure. *Procedia: Social and Behavioral Sciences*, *110*, 530–540. doi:10.1016/j.sbspro.2013.12.897

Molle, F. (2007). Scales and Power in River Basin Management: The Chao Phraya River in Thailand. *The Geographical Journal*, *173*(4), 358–373. doi:10.1111/j.1475-4959.2007.00255.x

Morales Pedraza, J. (2021). The Micro, Small, and Medium-Sized Enterprises and Its Role in the Economic Development of a Country. *Business and Management Research*, *10*(1), 33–44. doi:10.5430/bmr.v10n1p33

Morrar, R., Arman, H., & Mousa, S. (2017). The fourth industrial revolution (Industry 4.0): A social innovation perspective. *Technology Innovation Management Review*, *7*(11), 12–20. doi:10.22215/timreview/1117

Mudambi, S. M., & Schuff, D. (2010). What makes a helpful online review? A study of customer reviews on Amazon.com. *Management Information Systems Quarterly*, *34*(1), 185200. doi:10.2307/20721420

Muñoz, J. L. R., Ojeda, F. M., Jurado, D. L. A., Peña, P. F. P., Carranza, C. P. M., Berríos, H. Q., Molina, S. U., Farfan, A. R. M., Arias-Gonzáles, J. L., & Vasquez-Pauca, M. J. (2022). Systematic Review of Adaptive Learning Technology for Learning in Higher Education. *Eurasian Journal of Educational Research*, *98*(98), 221–233.

Murphy, C. (2009). *The International Organization for Standardization (ISO) global governance through voluntary consensus*. Routledge. doi:10.4324/9780203884348

Nakandala, D., Turpin, T., & Djeflat, A. (2015). Parallel innovation policies to support firms with heterogeneous innovation capabilities in developing economies. *Innovation and Development*, *5*(1), 131–145. doi:10.1080/2157930X.2014.980552

Narula, R., & Dunning, J. H. (2000). Industrial Development, Globalization, and Multinational Enterprises: New Realities for Developing Countries. *Oxford Development Studies*, *28*(2), 141–167. doi:10.1080/713688313

Nassar, N., Jafar, A., & Rahhal, Y. (2020). A novel deep multi-criteria collaborative filtering model for recommendation system. *Knowledge-Based Systems*, *187*, 104811. doi:10.1016/j.knosys.2019.06.019

Naumov, M. (2019). *On the dimensionality of embeddings for sparse features and data*. arXiv preprint arXiv:1901.02103.

Naveen Kumar, K., Prasad, G., Rajasekar, K., & Vadivelu, P. (2018, October). Satyanarayana Gollakota. Kavinprabhu S. K, A Study On The Forest Fire Detection Using Unmanned Aerial Vehicles. *International Journal of Mechanical and Production Engineering Research and Development*, *8*(7), 165–171.

Ng, K. H., Brady, Z., Ng, A. H., Soh, H. S., Chou, Y. H., & Varma, D. (2021). The status of radiation protection in medicine in the Asia-Pacific region. *Journal of Medical Imaging and Radiation Oncology*, *65*(4), 464–470. doi:10.1111/1754-9485.13165 PMID:33606359

Nichter, S., & Goldmark, L. (2009). Small Firm Growth in Developing Countries. *World Development*, *37*(9), 1453–1464. doi:10.1016/j.worlddev.2009.01.013

Norman, P. (2011). *The Risk Controllers: Central Counterparty Clearing in Globalized Financial Markets*. John Wiley & Sons.

Nowicki, M., Górski, Ł., & Bała, P. (2021). PCJ Java library as a solution to integrate HPC, Big Data, and Artificial Intelligence workloads. *Journal of Big Data*, *8*(1), 62. doi:10.118640537-021-00454-6

O'Bannon, C., Carr, J., Seekell, D. A., & D'Odorico, P. (2014). Globalization of agricultural pollution due to international trade. *Hydrology and Earth System Sciences*, *18*(2), 503–510. doi:10.5194/hess-18-503-2014

OECD. (2018). *The Future of Education and Skills: Education 2030*. OECD Publishing. https://www.oecd.org/education/2030/E2030%20Position%20Paper%20(05.04.2018).pdf

Oliver, R. L. (1997). A Conceptual Modes of Service Quality and Service Satisfaction: Compatible Goals, Different Concepts. *Advance in Services Marketing and Management, 2*, 65–85.

Onwudiwe, N. C., Tenenbaum, K., Boise, B. H., Elton, J., & Manning, M. (2018). *Real World Evidence: Implications and Challenges for Medical Product Communications in an Evolving Regulatory Landscape. Food and Drug Law Institute.*

OpenAI. (2023). *ChatGPT*. https://chat.openai.com/chat

Orlova, E. V. (2021). Design of Personal Trajectories for Employees' Professional Development in the Knowledge Society under Industry 5.0. *Social Sciences*, *10*(427). doi:10.3390ocsci10110427

Ostergaard, E. H. (2018). *Welcome to industry 5.0.* Retrieved from https://isajobs.isa.org/intech/20180403/

Ou, Y. C., & Verhoef, P. C. (2017). The impact of positive and negative emotions on loyalty intentions and their interactions with customer equity drivers. *Journal of Business Research*, *80*, 106–115. doi:10.1016/j.jbusres.2017.07.011

Oxford Learners Dictionaries. (2023). Skill. In *Oxford Learners Dictionaries*. https://www.oxfordlearnersdictionaries.com/definition/american_english/skill

Ozdemir, V., & Hekim, N. (2018). Birth of Industry 5.0: Making sense of big data with artificial intelligence, "the Internet of things" and next-generation technology policy. *OMICS: A Journal of Integrative Biology*, *22*(1), 65–76. doi:10.1089/omi.2017.0194 PMID:29293405

Pache, A.-C., & Chowdhury, I. (2012). Social Entrepreneurs as Institutionally Embedded Entrepreneurs: Toward a New Model of Social Entrepreneurship Education. *Academy of Management Learning & Education*, *11*(3), 494–510. doi:10.5465/amle.2011.0019

Palma, D., Ortúzar, J. D. D., Rizzi, L. I., Guevara, C. A., Casaubon, G., & Ma, H. (2016). Modelling choice when price is a cue for quality a case study with Chinese wine consumers. *Journal of Choice Modelling*, *19*, 24–39. doi:10.1016/j.jocm.2016.06.002

Parasuraman, A., Zeithaml, V.A. & Berry, L.L. (1988). SERVQUAL: A multiple-item scale for measuring consumer perceptions of service quality. *Journal of Retailing*, *64*, 12-40.

Parasuraman, A., Berry, L. L., & Zeithaml, V. A. (1993). More on improving service quality measurement. *Journal of Retailing*, *69*(1), 140–147. doi:10.1016/S0022-4359(05)80007-7

Park, J. Y., Back, R. M., Bufquin, D., & Shapoval, V. (2019). Servicescape, positive affect, satisfaction and behavioral intentions: The moderating role of familiarity. *International Journal of Hospitality Management, 78*, 102–111. doi:10.1016/j.ijhm.2018.11.003

Paschek, Mocan, & Draghici. (2019). *Industry 5.0 – The expected impact of the next industrial revolution.* https://www.researchgate.net/publication/336653504_The_Next_Industrial_Revolution_Industry_50_and_Discussions_on_Industry_40

PECB University. (2018). *ISO 21001:2018 – Educational organizations – Management systems for educational organizations – Requirements with guidance for use.* https://pecb.com/whitepaper/iso-210012018--educational-organizations--management-systems-for-educational-organizations--requirements-with-guidance-for-use

Peng, J., Biswas, A., Jiang, Q., Zhao, R., Hu, J., Hu, B., & Shi, Z. (2019). Estimating soil salinity from remote sensing and terrain data in southern Xinjiang Province, China. *Geoderma, 337*, 1309–1319. doi:10.1016/j.geoderma.2018.08.006

Pfeiffer, J., & Scholz, M. (2013). A Low-Effort Recommendation System with High Accuracy. *Business & Information Systems Engineering, 5*(6), 397–408. doi:10.100712599-013-0295-z

Pionline.com. (2018, November 12). Pension funds dominate largest asset owners. *Pion Line.* https://www.pionline.com/article/20181112/INTERACTIVE/181119971/pension-funds-dominate-largest-asset-owners

Powazek, D. M. (2002). *Design for Community: The Art of Connecting Real People in Virtual Places.* New Riders.

Pramanik, S., & Bandyopadhyay, S. K. (2014). Image Steganography Using Wavelet Transform and Genetic /Algorithm. *International Journal of Innovative Research in Advanced Engineering, 1*, 1–4.

Pramanik, S., Ghosh, R., Ghonge, M., Narayan, V., Sinha, M., Pandey, D., & Samanta, D. (2020). A Novel Approach using Steganography and Cryptography in Business Intelligence. In A. Azevedo & M. F. Santos (Eds.), *Integration Challenges for Analytics, Business Intelligence and Data Mining* (pp. 192–217). IGI Global.

Pramanik, S., & Suresh Raja, S. (2020). A Secured Image Steganography using Genetic Algorithm. *Advances in Mathematics: Scientific Journal, 9*(7), 4533–4541.

Prasad, G. (2020). Performance estimation of twin propeller in unmanned aerial vehicle. *INCAS Bulletin, 12*(2), 143–149. . doi:10.13111/2066-8201.2020.12.2.12

Prasad, G., Abhishek, P., & Karthick, R. (2018). Influence of Unmanned Aerial Vehicle in Medical Product Transport. *International Journal of Intelligent Unmanned Systems, 7*(2), 88-94. . doi:10.1108/IJIUS-05-2018-0015

Prasad, G., Vijayaganth, V., Sivaraj, G., Rajasekar, K., Ramesh, M., Raj, R. G., & Matheeswaran, P. (2018). Positioning of UAV using Algorithm for monitoring the forest region. *IEEE Xplore Digital Library*, 1361–1363. doi:10.1109/ICISC.2018.8399030

Prayag, G., Hassibi, S., & Nunkoo, R. (2019). A systematic review of consumer satisfaction studies in hospitality journals: Conceptual development, research approaches and future prospects. *Journal of Hospitality Marketing & Management*, *28*(1), 51–80. doi:10.1080/19368623.2018.1504367

QualityResearchInternational. (2022). Employability. In *Analytic Quality Glossary*. https://www.qualityresearchinternational.com/glossary/employability.htm#:~:text=Employability%20is%20the%20acquisition%20of,whether%20paid%20employment%20or%20not).&text=Employability%20usually%20refers%20to%20the,but%20this%20includes%20self%2Demployment.

Quelch, J. A., & Deshpande, R. (2004). *The Global Market: Developing a Strategy to Manage Across Borders*. John Wiley & Sons.

Radanliev, P., De Roure, D., Page, K., Nurse, J. R., Mantilla Montalvo, R., Santos, O., Maddox, L. T., & Burnap, P. (2020). Cyber risk at the edge: Current and future trends on cyber risk analytics and artificial intelligence in the industrial Internet of things and industry 4.0 supply chains. *Cybersecurity*, *3*(1), 1–21. doi:10.118642400-020-00052-8

Raja Sekar, K., Vignesh Moorthy, D., Prasad, G., Manigandan, P., & Senthil Kumar, M. (2018, October). An Experimental Investigation On The Influence Of Bio-Inspired Tubercle In The Unmanned Aerial Vehicle Propeller. *International Journal of Mechanical and Production Engineering Research and Development*, *8*(7), 404–409.

Randall, K. P., & Kroll, S. (2018). The customer is always right. *ABA Journal*, *104*(8), 30–32.

Rayan, R. A., Tsagkaris, C., & Iryna, R. B. (2021). The Internet of things for healthcare: applications, selected cases, and challenges. In *IoT in Healthcare and Ambient Assisted Living* (pp. 1–15). Springer. doi:10.1007/978-981-15-9897-5_1

Rey, N. (2016). *Combining UAV-imagery and machine learning for wildlife conservation*. Academic Press.

Reyes, D. R., van Heeren, H., Guha, S., Herbertson, L., Tzannis, A. P., Ducrée, J., Bissig, H., & Becker, H. (2021). Accelerating innovation and commercialization through standardization of microfluidic-based medical devices. *Lab on a Chip*, *21*(1), 9–21. doi:10.1039/D0LC00963F PMID:33289737

Richardson, W., & Mancabelli, R. (2011). *Personal learning networks: Using the power of connections to transform education*. Solution Tree Press.

Ritzer, G. (2008). *The Blackwell Companion to Globalization*. John Wiley & Sons.

Rocha, R. de R., Farazi, S., Khouri, R., & Pearce, D. (2011). The Status of Bank Lending to SMEs in the Middle East and North Africa Region: The Results of a Joint Survey of the Union of Arab Bank and the World Bank (*SSRN* Scholarly Paper ID 1794912). Social Science Research Network. https://papers.ssrn.com/abstract=1794912 doi:10.1596/1813-9450-5607

Rossi, B. (2018). *Manufacturing Gets Personal in Industry 5.0*. Reconteur.

Sachsenmeier, P. (2016). Industry 5.0—The Relevance and Implications of Bionics and Synthetic Biology. *Engineering (Beijing)*, *2*(2), 225–229. doi:10.1016/J.ENG.2016.02.015

Salameh, A., Hatamleh, A., Azim, M., & Kanaan, A. (2020). Customer oriented determinants of e-CRM success factors. *Uncertain Supply Chain Management*, *8*(4), 713–720. doi:10.5267/j.uscm.2020.8.001

Samanta, D., Dutta, S., Galety, M. G., & Pramanik, S. (2021). A Novel Approach for Web Mining Taxonomy for High-Performance Computing. *The 4th International Conference of Computer Science and Renewable Energies (ICCSRE'2021)*. 10.1051/e3sconf/202129701073

Sands, E. G., Bakthavachalam, V., & Reddick, R. (2020). *Global skills index 2020*. Academic Press.

Saniuk, S. G. S., & Straka, M. (2022). Identification of Social and Economic Expectations: Contextual Reasons for the Transformation Process of Industry 4.0 into the Industry 5.0 Concept. *Sustainability*, *14*(1391). doi:10.3390u14031391

Santa, R., MacDonald, J. B., & Ferrer, M. (2019). The role of trust in e-Government effectiveness, operational effectiveness and user satisfaction: Lessons from Saudi Arabia in e-G2B. *Government Information Quarterly*, *36*(1), 39–50. doi:10.1016/j.giq.2018.10.007

Saxena, A., Pant, D., Saxena, A., & Patel, C. (2020). Emergence of educators for industry 5.0: An Indological perspective. *International Journal of Innovative Technology and Exploring Engineering*, *9*(12), 359–363. doi:10.35940/ijitee.L7883.1091220

Schlie, E., & Yip, G. (2000). Regional follows global: Strategy mixes in the world automotive industry. *European Management Journal*, *18*(4), 343–354. doi:10.1016/S0263-2373(00)00019-0

Schot, J., & Steinmueller, W. E. (2018). Three frames for innovation policy: R&D, innovation systems, and transformative change. *Research Policy*, *47*(9), 1554–1567. doi:10.1016/j.respol.2018.08.011

Schuster, S. (2020, May 5). *Robotics and craftsmanship complement each other*. KUKA Blog.

Schwab. (2017). *The Fourth Industrial Revolution*. Crown Business.

Scriven, M. (1991). *Evaluation Thesaurus* (4th ed.). Sage.

Sergelen, D., Park, Y. S., & Lee, D. S. (2019). The Effect of Trust in Online Shopping Mall and Trust in Recommendation System on Cross-Purchase Intention. *The Articles of the Korean Industrial Engineering Society's Spring Joint Conference, 4*, 4821-4834.

Shahbazi, Z., & Byun, Y. C. (2019). Product recommendation based on content-based filtering using XGBoost classifier. *Int. J. Adv. Sci. Technol*, *29*, 6979–6988.

Shahbazi, Z., Hazra, D., Park, S., & Byun, Y. C. (2020). Toward improving the prediction accuracy of product recommendation system using extreme gradient boosting and encoding approaches. *Symmetry*, *12*(9), 1566. doi:10.3390ym12091566

Shah, S. F. H., Nazir, T., Zaman, K., & Shabir, M. (2013). Factors Affecting the Growth of Enterprises: A Survey of the Literature from the Perspective of Small- and Medium-Sized Enterprises. *Journal of Enterprise Transformation*, *3*(2), 53–75. doi:10.1080/19488289.2011.650282

Sharon, T. (2021). Blind-sided by privacy? Digital contact tracing, the Apple/Google API, and big tech's newfound role as global health policymakers. *Ethics and Information Technology*, *23*(S1), 45–57. doi:10.100710676-020-09547-x PMID:32837287

Sharu, H., & Guyo, D. W. (2013). *Factors Influencing Growth of Youth Owned Small and Medium Enterprises in Nairobi County, Kenya.*, *4*(4), 8.

Sheng, X., Li, J., & Zolfagharian, M. A. (2014). Consumer initial acceptance and continued use of recommendation agents: Literature review and proposed conceptual framework. *International Journal of Electronic Marketing and Retailing*, *6*(2), 112–127. doi:10.1504/IJEMR.2014.066467

Sherman, R. E., Anderson, S. A., Dal Pan, G. J., Gray, G. W., Gross, T., Hunter, N. L., LaVange, L., Marinac-Dabic, D., Marks, P. W., Robb, M. A., Shuren, J., Temple, R., Woodcock, J., Yue, L. Q., & Califf, R. M. (2016). Real-world evidence—What is it, and what can it tell us? *The New England Journal of Medicine*, *375*(23), 2293–2297. doi:10.1056/NEJMsb1609216 PMID:27959688

Sheth, J. N., & Parvatiyar, A. (2001). The antecedents and consequences of integrated global marketing. *International Marketing Review*, *18*(1), 16–29. doi:10.1108/02651330110381952

Sifakis, N. C., & Sougari, A.-M. (2003). Facing the Globalisation Challenge in the Realm of English Language Teaching. *Language and Education*, *17*(1), 59–71. doi:10.1080/09500780308666838

Singh, K., & Yip, G. S. (2000). Strategic Lessons from the Asian Crisis. *Long Range Planning*, *33*(5), 706–729. doi:10.1016/S0024-6301(00)00078-9

Sinha, M., Chacko, E., Makhija, P., & Pramanik, S. (2021). Energy Efficient Smart Cities with Green IoT. In Green Technological Innovation for Sustainable Smart Societies: Post Pandemic Era. Springer.

Skenderi, N., Islami, X., & Mulolli, E. (2017). The Impact of Informal Economy in the Development of SMEs-Evidence from Kosovo (2008-2012). *International Journal of Management, Accounting, and Economics*, *4*, 554–564.

SkillsFuture Singapore. (2022). *Skills Demand For the Future Economy*. https://www.skillsfuture.gov.sg/-/media/Skills-Report-2021/Skills-Report-Documents-FINAL/SSG-Skills_Demand_for_the_Future_Economy_2021.pdf

SkillsUSA. (2023). www.skillsusa.org

Smith, R. O., Scherer, M. J., Cooper, R., Bell, D., Hobbs, D. A., Pettersson, C., Seymour, N., Borg, J., Johnson, M. J., Lane, J. P., Sujatha, S., Rao, P. V. M., Obiedat, Q. M., MacLachlan, M., & Bauer, S. (2018). Assistive technology products: A position paper from the first global research, innovation, and education on assistive technology (GREAT) summit. *Disability and Rehabilitation. Assistive Technology*, *13*(5), 473–485. doi:10.1080/17483107.2018.1473895 PMID:29873268

Söderlund, M., & Dahlén, M. (2010). The "killer" ad: An assessment of advertising violence. *European Journal of Marketing*, *44*(11/12), 1811–1838. doi:10.1108/03090561011079891

Stahel, R. W., & MacArthur, E. (2019). *The Circular Economy: A User's Guide.* Taylor and Francis. https://www.taylorfrancis.com/books/mono/10.4324/9780429259203/circular-economy-walter-stahel-ellen-macarthur

Stergaard, E. H. (2016). *Industry 5.0 – Return of the human touch.* https://blog.universal-robots.com/industry-50-return-of-the-human-touch

Stern, A. D. (2017). Innovation under regulatory uncertainty: Evidence from medical technology. Journal of Public Economics, 145, 181–200.

Sun, Y., & Yannelis, N. C. (2007). Perfect competition in asymmetric information economies: Compatibility of efficiency and incentives. *Journal of Economic Theory*, *134*(1), 175–194. doi:10.1016/j.jet.2006.03.001

Suzuki, T., Tsuchiya, T., Suzuki, S., & Yamamba, A. (2016). Vegetation Classification Using a Small UAV Based on Superpixel Segmentation and Machine Learning. *Journal of The Remote Sensing Society of Japan*, *36*(2), 59–71.

Takiddin, A., Ismail, M., Nabil, M., Mahmoud, M. M., & Serpedin, E. (2020). Detecting electricity theft cyber-attacks in ami networks using deep vector embeddings. *IEEE Systems Journal*, *15*(3), 4189–4198. doi:10.1109/JSYST.2020.3030238

Tandon, U., Kiran, R., & Sah, A. (2017). Analyzing customer satisfaction: Users perspective towards online shopping. *Nankai Business Review International*, *8*(3), 266–288. doi:10.1108/NBRI-04-2016-0012

Tecuci, G., Boicu, M., Ayers, C., & Cammons, D. (2004). Cognitive assistants for analysts. In *The Proceedings of the 4th Annual Analysis & Productions's Analysis Congerence: The future of Analysis.* NSA. http://lac.gmu.edu/publications/2007/TecuciG_Cognitive_Assistants.pdf

Teo, H. S., Foerg-Wimmer, C., & Chew, P. L. M. (2016). *Medicines Regulatory Systems and Scope for Regulatory Harmonization in Southeast Asia.* Academic Press.

The Global Goals. (2015). *Goal #4 Quality Education.* https://www.globalgoals.org/goals/4-quality-education/?gclid=Cj0KCQjwiZqhBhCJARIsACHHEH-eh34l9GuxdCPCqItBp9mNLFiEozM9zWoGHnnun9PpMZv84ARV2hgaAoQ8EALw_wcB

Togo, M., & Gandidzanwa, C. P. (2021). The role of Education 5.0 in accelerating the implementation of SDGs and challenges encountered at the University of Zimbabwe. *International Journal of Sustainability in Higher Education*, 22(7), 1520–1535. doi:10.1108/IJSHE-05-2020-0158

Tonts, M., & Siddique, M. B. (2011). *Globalization, Agriculture and Development: Perspectives from the Asia-Pacific.* Globalization, Agriculture and Development. https://www.elgaronline.com/view/edcoll/9781847208187/9781847208187.00005.xml

Treiblmaier, H., & Sillaber, C. (2021). The impact of blockchain on e-commerce: A framework for salient research topics. *Electronic Commerce Research and Applications*, 48, 101054. doi:10.1016/j.elerap.2021.101054

Tridico, P., & Paternesi Meloni, W. (2018). Economic growth, welfare models, and inequality in the context of globalization. *Economic and Labour Relations Review*, 29(1), 118–139. doi:10.1177/1035304618758941

U20 Rome-Milan. (2021, June 17). *Urban 20 calls on G20 to empower cities to ensure a green and just recovery.* Urban 20. https://www.urban20.org/wp-content/uploads/2021/06/U20-2021-Communique-Final.pdf

United Nations - Addis Ababa Action Agenda. (2015). *Addis Ababa Action Agenda of the Third International Conference on Financing for Development.* UN. https://www.un.org/esa/ffd/wp-content/uploads/2015/08/AAAA_Outcome.pdf

United Nations. (2015). *Sustainable Development Goals.* https://www.un.org/sustainabledevelopment/blog/2015/12/sustainable-development-goals-kick-off-with-start-of-new-year/

United Nations. (2015). *The 17 Goals.* UN. https://sdgs.un.org/goals

Urata, S. (2002). Globalization and the Growth in Free Trade Agreements. *Asia-Pacific Review*, 9(1), 20–32. doi:10.1080/13439000220141569

Uskov, V. L., Bakken, J. P., Gayke, K., Fatima, J., Galloway, B., Ganapathi, K. S., & Jose, D. (2020). *Smart Learning Analytics: Student Academic Performance Data Representation, Processing and Prediction.* Springer Singapore. doi:10.1007/978-981-15-5584-8_1

Varadarajan Sowmya, D., Majumdar, S., & Gallant, M. (2010). Relevance of education for potential entrepreneurs: An international investigation. *Journal of Small Business and Enterprise Development*, 17(4), 626–640. doi:10.1108/14626001011088769

Velasquez, M. (2000). Globalization and the Failure of Ethics. *Business Ethics Quarterly*, 10(1), 343–352. doi:10.2307/3857719

Venkatesh, V., Morris, M. G., Davis, G. B., & Davis, F. D. (2003). User acceptance of information technology: Toward a unified view. *Management Information Systems Quarterly*, 27(3), 425–478. doi:10.2307/30036540

Vujakovic, P. (2009). *How to Measure Globalization? A New Globalisation Index (NGI) (Working Paper No. 343).* WIFO Working Papers. https://www.econstor.eu/handle/10419/128904

Wagih, H. M., & Mokhtar, H. M. O. (2021). Ridology: An Ontology Model for Exploring Human Behavior Trajectories in Ridesharing Applications. In M. Al-Emran, K. Shaalan, & A. Hassanien (Eds.), *Recent Advances in Intelligent Systems and Smart Applications. Studies in Systems, Decision and Control* (Vol. 295). Springer. doi:10.1007/978-3-030-47411-9_30

Wagner, I. (2019). Projected commercial drone revenue worldwide 2016-2025. Statista GmbH.

Walia, N., Srite, M., & Huddleston, W. (2016). Eyeing the web interface: The influence of price, product, and personal involvement. *Electronic Commerce Research*, *16*(3), 297–333. doi:10.100710660-015-9200-9

Wang, W., & Wang, M. (2019). Effects of sponsorship disclosure on perceived integrity of biased recommendation agents: Psychological contract violation and knowledge-based trust perspectives. *Information Systems Research*, *30*(2), 507–522. doi:10.1287/isre.2018.0811

Wang, Y., Han, L., Qian, Q., Xia, J., & Li, J. (2022). Personalized recommendation via multi-dimensional meta-paths temporal graph probabilistic spreading. *Information Processing & Management*, *59*(1), 102787. doi:10.1016/j.ipm.2021.102787

Wang, Y., Tang, T., & Tang, J. (2001). An instrument for measuring customer satisfaction toward web sites that market digital products and services. *Journal of Electronic Commerce Research*, *2*(3), 89–102.

Warhuus, J. P., Blenker, P., & Elmholdt, S. T. (2018). Feedback and assessment in higher-education, practice-based entrepreneurship courses: How can we build legitimacy? *Industry and Higher Education*, *32*(1), 23–32. doi:10.1177/0950422217750795

Watson, J. (2003). The potential impact of accessing advice on SME failure rates: The potential impact of accessing advice on SME failure rates. *Proceedings of the Small Enterprise Association of Australia and New Zealand 16th Annual Conference.* SEA.

Weill, P., & Woerner, S. L. (2018). *What's Your Digital Business Model? Six questions to help you build the next-generation enterprise.* Harvard Business Review Press.

Welch, N. (2022). *Is humanity at the centre of your leadership and team culture?* Natalie Welch. https://www.nataliewelch.com.au/is-humanity-at-the-centre-of-your-leadership-and-team-culture/

Wen, J., Zhu, X. R., Wang, C. D., & Tian, Z. (2022). A framework for personalized recommendation with conditional generative adversarial networks. *Knowledge and Information Systems*, *64*(10), 2637–2660. Advance online publication. doi:10.100710115-022-01719-z

White, R. D. (2010). *Global Environmental Harm: Criminological Perspectives.* Taylor & Francis.

Williams, D., Hricko, M., & Howell, S. L. (Eds.). (2006). *Online Assessment, Measurement and Evaluation: Emerging Practices. IGI Global.* IGI Global. doi:10.4018/978-1-59140-747-8

World Health Organization. (2013). *Medical devices and eHealth solutions: Compendium of innovative health technologies for low-resource settings 2011-2012*. World Health Organization.

World Health Organization. (2017). *China's policies promote the local production of pharmaceutical products and protect public health*. World Health Organization.

World Health Organization. (2020). Rational use of personal protective equipment (PPE) for coronavirus disease (COVID-19): Interim guidance, 19 March 2020 (No. WHO/2019-Nov/IPC PPE_use/2020.2). World Health Organization.

World Trade Organization. (2016). *World Trade Report 2016: Levelling the trading field for SMEs*. WTO. https://www.wto.org/english/res_e/booksp_e/world_trade_report16_e.pdf

Wu, H. (2021, February). Application of Collaborative Filtering Personalized Recommendation Algorithms to Website Navigation. *Journal of Physics: Conference Series*, *1813*(1), 012048. doi:10.1088/1742-6596/1813/1/012048

Xiao, B., & Benbasat, I. (2014). Research on the Use, Characteristics, and Impact of e-Commerce Product Recommendation Agents: A Review and Update for 2007-2012. In Handbook of Strategic eBusiness Management (pp. 403-431). Springer Berlin Heidelberg.

Xiao, B., & Benbasat, I. (2015). Designing Warning Messages for Detecting Biased Online Product Recommendations: An Empirical Investigation. *Information Systems Research*, *26*(4), 793–811. doi:10.1287/isre.2015.0592

Xiao, B., & Benbasat, I. (2018). An empirical examination of the influence of biased personalized product recommendations on consumers' decision making outcomes. *Decision Support Systems*, *110*, 46–57. doi:10.1016/j.dss.2018.03.005

Xie, S., Zhang, F., & Cheng, R. (2021). Security Enhanced RFID Authentication Protocols for Healthcare Environment. *Wireless Personal Communications*, *117*(1), 71–86. doi:10.100711277-020-07042-6

Xu, J., Benbasat, I., & Cenfetelli, R. T. (2014). The Nature and Consequences of Trade-off Transparency in the Context of Recommendation Agents1. *Management Information Systems Quarterly*, *38*(2), 379–406. doi:10.25300/MISQ/2014/38.2.03

Yameogo, C. E. W., Omojolaibi, J. A., & Dauda, R. O. S. (2021). Economic globalization, institutions and environmental quality in Sub-Saharan Africa. *Research in Globalization*, *3*, 100035. doi:10.1016/j.resglo.2020.100035

Yi, C., Jiang, Z., & Benbasat, I. (2017). Designing for diagnosticity and serendipity: An investigation of social product-search mechanisms. *Information Systems Research, 28*(2), 413-429.

Yip George, S. (2004). Using Strategy to Change Your Business Model. *Business Strategy Review*, *15*(2), 17–24. doi:10.1111/j.0955-6419.2004.00308.x

Yip, G. S., Biscarri, J. G., & Monti, J. A. (2000). The Role of the Internationalization Process in the Performance of Newly Internationalizing Firms. *Journal of International Marketing*, *8*(3), 10–35. doi:10.1509/jimk.8.3.10.19635

Yip, G. S., Johansson, J. K., & Roos, J. (1997). Effects of Nationality on Global Strategy. *MIR. Management International Review*, *37*(4), 365–385.

Yiu, C.S., Grant, K., & Edgar, D. (2007). Factors affecting the adoption of internet banking in Hong Kong – implications for the banking sector. *International Journal of Information Management*, *27*(5), 336–351.

You, J.-I. (1999). Income Distribution and Growth in East Asia. In *East Asian Development: New Perspectives*. Routledge.

Yuce, B., Ghalaty, N. F., Santapuri, H., Deshpande, C., Patrick, C., & Schaumont, P. (2016). Software fault resistance is futile: Effective single-glitch attacks. Paper presented at the *2016 Workshop on Fault Diagnosis and Tolerance in Cryptography (FDTC)*. IEEE. 10.1109/FDTC.2016.21

Zaman, B., Mac Mckee, D., & Jensen, A. (2017). *UAV, Machine Learning*. And GIS for Wetland Mitigation in Southwestern Utah.

Zenkour, A. M., & El-Shahrany, H. D. (2021). Hygrothermal forced vibration of a viscoelastic laminated plate with magnetostrictive actuators resting on viscoelastic foundations. *International Journal of Mechanics and Materials in Design*, *17*(2), 301–320. doi:10.100710999-020-09526-6

Zhang, H., Wang, Z., Chen, S., & Guo, C. (2019). Product recommendation in online social networking communities: An empirical study of antecedents and a mediator. *Information & Management*, *56*(2), 185–195. doi:10.1016/j.im.2018.05.001

Zhang, L., Lam, W., & Hu, H. (2013). A case study of leading medical device companies in China is a complex product and system, catch-up, and sectoral system of innovation. *International Journal of Technological Learning, Innovation and Development*, *6*(3), 283–302.

Zhang, M., Lettice, F., & Zhao, X. (2015). The impact of social capital on mass customization and product innovation capabilities. *International Journal of Production Research*, *53*(17), 5251–5264. doi:10.1080/00207543.2015.1015753

About the Contributors

Mahmoud Numan Bakkar has a Ph.D. in Information Technology/Computer Engineering and Business Information Systems from RMIT University, Australia. He has a Master's degree in Information Technology Management from the University of Wollongong (UOW), Australia, and a Bachelor of Computer Engineering, from Jordan University of Science and Technology (JUST), Jordan. He has extensive experience in teaching and training in the UAE and Australia. Furthermore, he taught courses for both undergraduate and postgraduate levels in Australia and the UAE. Also, he has a vast knowledge of business development, management, and leadership. Dr. Mahmoud Bakkar's current research interests include Artificial Intelligence Systems, Industry 4.0 applications, Industry 5.0 applications, Smart city systems and innovations, Cloud Computing; Cybersecurity, IT in Education, Assessment Measurement; Data Mining; Big Data and Business Intelligence Analytics; Cognitive Computing and their application in Business; IT Innovation & Management; Instructional Design, Online Training and Educations, E-Commerce, and Online Business Development.

Elspeth McKay, PhD, is Director of Cogniware.com.au since her change of career as an Associate Professor of Information Systems (IS) at the RMIT University, School of Business IT and Logistics, Melbourne, Australia. She is passionate about learning analytics (psychometric testing) and designing effective eLearning resources for the education sector and industry training/reskilling programmes, including investigations of how individuals interpret text and graphics within web-mediated learning environments. She has designed e-Learning tools implemented through rich internet applications including ARPS – an advanced repurposing pilot system, COGNIWARE – a multi-modal e-Learning framework, GEMS – a global eMuseum System, eWRAP – Electronic work readiness awareness programme, EASY – Educational/academic (skills) screening for the young, offering enhanced accessibility through touch screen technologies. Over the last decade Dr McKay has published extensively in the research fields of HCI and educational technology. In recognition of her contribution to the professional practice of IS research, she was elected as a Fellow of the Australian Computer Society (FACS).

* * *

Prasad G. is currently working as faculty in the department of Aerospace Engineering Chandigarh University, India. He has 8 years of experience in Teaching and research. He received his Bachelor of Engineering and Master of Engineering degrees with first class and distinction in the field of aeronautical Engineering from the Anna University, Chennai, India. Prof. Prasad G is the author of over 30 technical publications, proceedings, editorials and books. His research interests include Unmanned Aerial Vehicles, Computational Fluid Dynamics and Interdisciplinary research. He has completed two funded project sponsored by The Institution of Engineers (India) and Tamilnadu State Council for Science and Technology. Awarded Indian National Science Academy (INSA) Visiting Scientist Programme 2019 and Awarded Science Academies' Summer Research Fellowship Programme (SRPF) 2019. He is a Professional Membership of American Institute of Aeronautics and Astronautics, Institution of Engineers, Life Member in Indian Cryogenic Council and Life Member in Shock Wave Society. Reviewer in Aircraft Engineering and Aerospace Technology, International Journal of Engine Research and Journal of The Institution of Engineers (India): Series C.

Sunil Gupta has over more than 19 Years of experience in teaching and research in the field of Computer science and Engineering. He is working as a Professor at the University of petroleum and energy studies (UPES). He is an Associate Member, Computer Society of India, Member, Computer Science Teacher Association, Life Member, International Association of Engineers, Member, International Association of Computer Science and Information Technology, Member, Internet Society (ISOC), and IEEE Society.

Monit Kapoor is a focused individual with a flair for teaching/faculty/mentoring roles and an active desire for research in the area of Computer Engineering. My objectives are to do well in the area of University Level Teaching/Research Positions. I also look forward to working with Industry in the areas of DevOps, Cloud Computing Solutions, IoT, Wireless Networks, and Artificial Intelligence on joint projects/Collaborative Research. I also offer technical sessions in webinar mode or skype mode

Arshia Kaul is an Assistant Professor of Operations Management and Research with 10+ years of rich work experience across the teaching and research fields. Currently she is the Consultant Carpediem EdPsych Consultancy. She has worked at ASMSOC, NMIMS University, Mumbai, India (2019-2022). She has earlier worked as a research fellow at IFCPAR, CSIR-NISTADS, India on various different projects. She has published 20+ research papers in journals of national and international repute and presented at 10+ conferences. One of her most recent articles was published in Business Strategy and the Environment (Impact factor- 10.302).

Rita Komalasari lectures at the Faculty of Medicine and Graduate School, YARSI University. The author actively writes scientific articles and is published in national and international journals. Her current work is focused on the impact of regulatory changes on Healthcare applications in Industry 5.0 and its capacity to develop and deliver affordable healthcare products. The author is also a reviewer in several national and international journals. Orchid ID: , Scholar ID: , Scopus ID: https://www.scopus.com/authid/detail.uri?authorId=57219652348.e805af06-7543-4bbf-9f00-4fc4b0155044

Siddharth Misra is an eminent and thorough researcher in social science, research, and development. He has published than 40 publications with A, B, and C category publications, Scopus indexed and Web of Science publications to his credit. He is an expert in HR analytics and qualitative research.

Mohammad Izzuddin Mohammed Jamil is currently a PhD Candidate at Universiti Brunei Darussalam, conducting researches in the field of Business Management and Entrepreneurship. He received his Bachelor's Degree in Business Administration at Coventry University, United Kingdom, and Master's Degree in Management at Universiti Brunei Darussalam. He has also studied other courses including Software Engineering and Programming, Multimedia and Graphic Design. At Universiti Islam Sultan Sharif Ali, he studied Intermediate Level Arabic Language, Tadabbur Al-Quran programme and Pengurusan Jenazah Mazhab Syafi'i course. His industry experiences started when he first became a salesman and distributor, and has since worked in various industries including becoming a Tutor/Teacher, App Developer, Graphic Designer and Business Executive Officer. He is teaching in various academic branches in Entrepreneurship, Business Management, Financial and Management Accounting, Macroeconomics and Microeconomics, Commerce, and Information and Communication Technology. He has also been involved in academic research for several years, and his research areas include Entrepreneurship, Growth and MSMEs.

Parameswar Nayak is an ardent Professor in Human Resource Management, an Academic Administrator, Corporate Trainer, and Management Consultant. He has about 33 years of work experience in teaching (in India and overseas), academic administration, training, research and consultancy, and 20 years in leadership positions. He has published three books and 30 research papers in Scopus indexed and other peer-reviewed journals besides his Doctoral guidance.

Priyadarsini Patnaik is a knowledgeable, experienced, and seasoned marketing professional with a huge experience in both industry and academics. Currently pursuing Ph.D. in Birla Global University, Bhubaneswar, Odisha, India. Her research area of interest are: Artificial Intelligence, Consumer Behaviour, Advertising, Sales & Distribution Management, Digital Marketing, Rural consumer behaviour.

Guillermo Pivetta has experience as a senior managing and strategy consultant and head researcher for international think tanks. Since 2017 he has played roles in G20 events during the Argentine, Japanese, Kingdom of Saudi Arabia, Italia, Indonesia, and India presidencies. He has also been active in T20, Urban20 and Youth20 projects. Currently he is a member of the B20 Digitalization Task Force, one of the G20 high level advisory boards. Guillermo has a broad based set of interests from innovation management, digital democracy and blockchain and the future cooperative economy; while working at high level projects for new data and analytics and insights for cities and investors, the role of trust in the new economy, and the changing economic management of the world of finance for consumers and business. Guillermo having a background involving entrepreneurial start-ups in the digital field and software development, has combined these with his intense interest in education and psychology having held positions as Associate Professor in the National Technological University of the Argentine. He holds three specializations degrees -Education Management (UdeSA), Human Development (FLACSO) and Innovation Management (UBA), with a BSc. degree in Psychology (UNMDP). He remains a creative leader in all activities that he engages with, and is a passionate champion for a transition to a better future global economy.

Sabyasachi Pramanik is a Professional IEEE member. He obtained a PhD in Computer Science and Engineering from the Sri Satya Sai University of Technology and Medical Sciences, Bhopal, India. Presently, he is an Associate Professor, Department of Computer Science and Engineering, Haldia Institute of Technology, India. He has many publications in various reputed international conferences, journals, and online book chapter contributions (Indexed by SCIE, Scopus, ESCI, etc.). He is doing research in the field of Artificial Intelligence, Data Privacy, Cybersecurity, Network Security, and Machine Learning. He is also serving as the editorial board member of many international journals. He is a reviewer of journal articles from IEEE, Springer, Elsevier, Inderscience, IET, and IGI Global. He has reviewed many conference papers, has been a keynote speaker, session chair and has been a technical program committee member in many international conferences. He has authored a book on Wireless Sensor Network. Currently, he is editing 6 books from IGI Global, CRC Press EAI/Springer and Scrivener-Wiley Publications.

Lila Rajabion is Assistant Professor and coordinator of the Master of Science in Information Technology (MSIT) program at SUNY Empire State College, where she teaches and develops the MSIT curriculum with a concentration in cybersecurity and Web design. She has more than 20 years' experience conducting research and providing consulting in various dimensions of IT combined in the academia and private sectors. Dr. Rajabion also has significant professional experience in providing leadership in the areas of systems analysis and design, cybersecurity, enterprise software application development, and IT project management for local and global projects. In addition, she has conducted various needs-based training programs. Dr. Rajabion has written many publications and has participated in research grants for the National Center for Women & Information Technology, which focuses on women in STEM. She is a long-time advocate for increasing participation and retention of women and minorities in the IT workforce. Dr. Rajabion has worked on many projects in this domain.

Hitesh Kumar Sharma is an Associate Professor in the department of cybernetics, SoCS, University of Petroleum & Energy Studies. He has published more than 50 Research papers in International and National Journals. He has authored two books. He has also filed and published 02 Patents. His research area is Network security and Machine Learning.

Reymond Voutier is a seasoned international business leader with decades of experience in project research, development, and marketing. His career began in product design and he founded Design Audit, a successful design consultancy that was ultimately acquired by Litton Industries, then the world's largest business conglomerate. Mr Voutier also worked as a management consultant for major clients across a range of industries. In recent years, he has been spearheading "Navigating the Future," an initiative focused on AI and data assessment tools and how they can be leveraged by Cities, Business Enterprises and citizens to gain a better understanding of the Future of Work and Education issues. He has established strong relationships with City associations like Metropolis, the Global Parliament of Mayors and the OECD Champion Mayors, among others. Mr Voutier's expertise and extensive experience have earned him a respected reputation within the international business community. He is a member of several prominent G20 task forces, including groups within the B20, and T20. He is also a member of the Council on African Economic Integration.

Index

Ingram Content Group UK Ltd.
Milton Keynes UK
UKHW052146080623
423139UK00007B/194